C. O. Weiss, R. Vilaseca   **Dynamics of Lasers**

Φ Published in collaboration with
Deutsche Physikalische Gesellschaft

Nonlinear Systems
Concepts – Methods – Applications

Edited by H.-G. Schuster

This new series is intended to provide expert information on: new theoretical concepts (e. g., strange attractors, cellular automata, neuronal networks), recently developed experimental methods (e. g., for the on-line analysis of chaotic data), possible applications (e. g., for semiconductor devices, photochemistry, modelling of physical, chemical or biological systems, computer science).

Volume 1 Dynamics of Lasers
by C. O. Weiss and R. Vilaseca

© VCH Verlagsgesellschaft mbH, D-6940 Weinheim (Federal Republic of Germany), 1991

Distribution
VCH, P. O. Box 101161, D-6940 Weinheim (Federal Republic of Germany)
Switzerland: VCH, P. O. Box, CH-4020 Basel (Switzerland)
United Kingdom and Ireland: VCH, Wellington Court, Wellington Street, Cambridge CB1 1HZ (England)
USA and Canada: VCH, Suite 909, 220 East 23rd Street, New York, NY 10010–4606 (USA)

ISBN 3-527-26586-4 (VCH, Weinheim)   ISBN 0-89573-966-6 (VCH, New York)

C. O. Weiss, R. Vilaseca

Dynamics of Lasers

Weinheim · New York · Basel · Cambridge

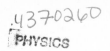
.4370260

Authors:
Dr. C. O. Weiss
Physikalisch - Technische
Bundesanstalt
Bundesallee 100
D-3300 Braunschweig

Prof. R. Vilaseca
Facultat de Fisica
Universitat de Valencia
Dr. Moliner, 50
E-46100 Burjassot (Valencia)

Editor:
Prof. Dr. H. G. Schuster
Inst. für Theoretische Physik und
Sternwarte der Universität
Olshausenstraße 40
D-2300 Kiel 1

Published jointly by
VCH Verlagsgesellschaft mbH, Weinheim (Federal Republic of Germany)
VCH Publishers, Inc., New York, NY (USA)

Editorial Director: Walter Greulich
Production Manager: Dipl.-Ing. (FH) Hans Jörg Maier

Library of Congress Card No. applied for

British Library
Cataloguing-in-Publication Data
Weiss, C. O.
    Dynamics of lasers.
    1. Lasers
    I. Title II. Vilaseca, R.
    III. Schuster, H. G.
    621.366
    ISBN 3-527-26586-4

Deutsche Bibliothek Cataloguing-in-Publication Data:
    Die Deutsche Bibliothek - CIP-Einheitsaufnahme
    Weiss, Carl O.:
    Dynamics of lasers / C. O. Weiss; R. Vilaseca. -
    Weinheim; New York; Basel; Cambridge: VCH, 1991
    (Nonlinear systems; Vol. 1)
    ISBN 3-527-26586-4 (Weinheim ...)
    ISBN 0-89573-966-6 (New York)
    NE: Vilaseca, Ramon:; GT

Composition and Printing: Krebs-Gehlen, D-6944 Hemsbach.
Bookbinding: Wilh. Osswald + Co · Großbuchbinderei, D-6730 Neustadt/Weinstraße
Printed in the Federal Republic of Germany

Dedication

To my family (Irene, David and Roger), who encouraged me to write this book and patiently supported my intensive dedication to this "homework" (R. V.)

To Jutta, who pushed me to continue working (C. O. W.)

Preface

It has become evident over the last few years that dissipative systems through which energy is pumped are not usually in a time-independent state. In generall they are in a dynamic state, and the time-dependent change of their variables can be periodic, multiperiodic, or deterministically chaotic. Lasers - typical "pumped" dissipative systems — are no exception. Fascinating in their property of spontaneous phase-symmetry breaking and spontaneous formation of well-organized coherent field structures at the laser threshold, they also constitute in the simplest case, a beautiful realization of *the* paradigm of nonlinear dynamics, the Lorenz model.

The linking of lasers and fluids by the Lorenz model goes beyond this basic case. For examples, the existence of field "defects" in laser emission has recently been shown, and their equivalence with fluid vortices has been established. It is therefore not surprising to observe that lasers exhibit a multitude of dynamical phenomena, some even resembling fluid turbulence.

It is found in many cases that the intrinsic dynamics of lasers is at the root of unexplainable laser properties or is overlooked in laser applications. In fact, the dynamics of lasers often interferes with specific applications, and a better understanding of this phenomenos may prevent such problems from occurring. Clearly, however, dynamics can also be used to one's advantage. For instance, it is possible to generate particular types of coherent radiation, such as laser emission with energy compressed into coherent pulses or radiation with well-characterized incoherences.

This book is intended to familiarize those interested in nonlinear dynamics in general, in laser physics or nonlinear optics, or in laser applications with the dynamical properties of lasers. Since an intuitive understanding of dynamical phenomena is most easily attained using simple, general concepts of nonlinear dynamics - the most prominent among them being fixed points — we have given substantial room to these principles and to two typical examples. The emphasis is not on mathematics but on intuitive concepts. The cases treated and discussed concern situations in which all kinds of lasers are involved, from the academic case of the radio frequency NMR ruby "laser" to semiconductor diode lasers under conditions which are typical of technical laser applications.

The authors would like to thank the following organizations and individuals:

Physikalisch-Technische Bundesanstalt and Deutsche Forschungsgemeinschaft as well as the IFIC computer center and DGICYT for their support; F. Silva for the

<blockquote>
</blockquote>

numerical calculation and plotting of many of the original figures in the book; R. Corbalán for many fruitful discussions and for benefical criticism of several chapters; C. Simó for some useful explanations regarding mathematical aspects of nonlinear dynamics; J. Pujol and F. Laguarta for their useful input on three-level lasers; L. Roso for the reading of some of the chapters; M. Tietzek and F. Pérez for the preparation and organization of the manuscript; W. Fellner for the drawing of a large part of the figures.

We also thank Prof. H. G. Schuster for the suggestion to write this book and for a critical reading of the manuscript as well as Dr. W. Greulich for his editorial support.

C. O. Weiß, R. Vilaseca   Braunschweig, Valencia, January 1991

Contents

Introduction

In this book we treat the dynamic properties of lasers. Interest in this field is only a few years old, the stable continuously emitting laser having been at the center of attention for the last 25 years.

This interest is to a great extent motivated by the analogies of laser dynamics with dynamic, chaotic and turbulent processes in other fields of nature. On the other hand, it has become increasingly evident that ununderstood and "troublesome" laser properties such as unstable emission, poor reproducibility of laser pulses, limitations of attainable widths of ultrashort pulses and coherence lengths, and also problems with the emission mode pattern shapes, spontaneous irregular pulsing, etc., are not caused by insufficient technical skill, but are direct consequences of the inherent non-linearities of the laser. The improvement of these laser properties therefore requires not just dilligent effort but more an understanding of the physics of these particular non-linear dissipative systems.

Apart from assisting in this respect, an understanding of laser dynamics has helped in interpreting the rich phenomena of particular systems, that have been known for a long time. A good example is the laser with an internal saturable absorber. In addition, it can lead to new useful applications, such as the generation of phase coherent pulses in the infrared region, which combines the advantage of temporal compression of an average power into high-power short pulses with the coherence of continuous lasers.

Dynamics of lasers can be caused by the intrinsic properties of the laser systems, as in self-pulsing, self-Q-switching lasers or through external influences on the laser, such as time dependence of laser parameters, or interaction with external optical fields. The book excludes the case of mode-locking of multi-longitudinal modes in a laser. This case, as it is of considerable practical importance, is the subject of many other books devoted specifically to it. Apart from the more practical questions of the conditions under which a particular laser system is stable, i.e. showing steady-state emission, or the conditions that will produce pulses in a possibly desired manner, which are covered in this book, the dynamics of lasers is treated in the more general context of non-linear dynamics, as it appears that today laser physics, like hardly any other field in physics, lends itself to experiments that permit questions of non-linear dynamics in general to be explored.

For this reason, one of the most basic theoretical models of non-linear dynamics (and fluid dynamics) — the Lorenz model — which happens also to describe the simplest case of a laser, is discussed in some detail as a typical example, together with the logistic map, the dynamics of which is frequently encountered in nature and also with lasers. These two examples give a taste of the mathematical treatment of non-linear dynamics problems either through maps (discrete dynamics) or through differential equations (time-continuous dynamics). The mathematical concepts of the treatment of non-linear dynamics problems are introduced, but not in a rigid mathematical way, rather in intuitive terms, intended to make the reader familiar with the non-linear dynamics literature. If the reader is not interested in this aspect, it is possible to skip over these chapters and look up technical terms later when necessary for the understanding of the systems described.

The book is organized as follows:

After an introductory Chapter 1 which treats the historic roots of laser dynamics first and then gives an overview of the properties of different kinds of lasers, two basic cases relating to laser physics are treated in Chapter 2: the Lorenz model and the logistic equation, illustrating the two common ways of treating dynamical problems, i.e. time continuous and discrete, while introducing terms. Concepts of the mathematics governing non-linear dynamics in general are covered in Chapter 3, and Chapter 4 shows the concepts and methods by which dynamic behavior can be characterized and quantified.

The laser cases described in the following chapters are arranged according to their complexity. In Chapter 5 we start with the simplest case: the homogeneously broadened, traveling wave single-mode laser. This is then extended to inhomogeneous broadening. The first case is exemplified by gas lasers with optical excitation and the second by high-gain He − Xe and He − Ne lasers. The next more complex case is lasers with a saturable absorber inside the resonator. We treat gas lasers and semiconductor diode lasers as typical cases in Chapter 6.

Lasers that are not capable of showing dynamics by themselves, but do so on adding degrees of freedom such as a periodic modulation of one laser parameter (gain, resonator loss, resonator frequency, medium line width) or by an external feedback path, be it electronic or purely optical, or when subjected to an external optical field, are then described. Gas lasers, solid-state lasers and a more exotic laser species, the radio-frequency nuclear magnetic resonance "laser", serve as examples (Chapter 7). The example of feedback on diode lasers illustrates limitations of stabilizing the laser output, e.g. its frequency, by dynamic instabilities.

Chapter 8 is devoted to gas lasers with laser excitation, because this class of lasers, emitting from the millimeter wavelength range to the visible spectrum, show a particularly rich variety of dynamic behavior.

Chapter 9 treats selected cases of multimode lasers and Chapter 10 extends the phenomena to spatio-temporal dynamics and instabilities , with inclusion of transverse variations of laser parameters and variables. This last field is still in its infancy but promises, in addition to interesting new physics, useful applications in optical computing, pattern recognition and possibly non-classical states of radiation.

1 Historical Background; Properties of Lasers

1.1 Dynamic Behavior of Lasers

As said in the Introduction, a laser can have different forms of emission as far as the time dependence is concerned. The output power and laser frequency may be constant or variable in time. For many applications, the constant or "continuous wave" (cw) emission is the preferred mode.

A laser consists of a medium that is "pumped" and can amplify light. In order to build an optical oscillator from this optical amplifier, some of the amplified light is taken from the output of the amplifier and fed back, usually by utilizing mirrors, into the amplifier input, so that a constant emission takes place. This scheme is most obviously realized in the "ring" laser geometry (Fig. 1.1.1). This kind of apparatus produces a traveling wave optical field within the "resonator" formed by the three mirrors. The mirror arrangement is called a "resonator" because at certain wavelengths the interference of the waves traveling around the ring multiple times is constructive everywhere. At these wavelenghts, a high field can therefore build up inside the reso- nator. The condition for constructive interference is of course that the ring perimeter l be a multiple of the optical wavelength λ. Since the dimension l of the ring is usually

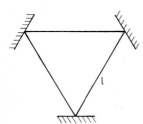

Fig. 1.1.1. "Ring" resonator consisting of three mirrors. The ring perimeter length is l.

long in comparison with λ, the optical resonator has many wavelengths for which interference is constructive or the resonator is "resonant". This can lead to laser emission at several wavelengths simultaneously.

A simpler way of providing the optical feedback is to re-inject some of the amplified light back into the "output" of the amplifier. Since the gain medium has no preferred direction in general, the input and output are indistinguishable and the mirror arrangement can be made simpler (Fig. 1.1.2).

The difference with the ring geometry is that in this linear resonator the interference varies spatially along the axis periodically from constructive to destructive. Thus the linear resonator produces a standing wave field with maxima and "zeros" of the field amplitude. The conditions for constructive interference in the linear case is $l = n \cdot \lambda/2$, where n is an integer. Again, since $l \gg \lambda$ there are many "resonances".

Since the field distribution of the ring resonator is so much simpler (constant along the axis) than that of the linear resonator, the ring is the preferred type for theoretical models, while the simplicity of the linear resonator makes it the preferred choice in most practical cases.

In many applications and in most theoretical work, a "single-mode" laser is the preferred type. In practice, this requires additional optical elements in the resonator or outside, which favor one particular resonance (mode selector), e.g. additional smaller interferometers.

For the dynamic behavior of a laser, it is often important that the latter can have two reservoirs of energy. The energy may be contained in the laser field or in the medium. Thus the laser resembles a pendulum or LC circuit having a pronounced resonance frequency. The resonance may be more or less strongly damped. If it is relatively weakly damped, the resonance manifests itself, e.g. when the laser is rapidly switched on.

Figure 1.1.3 shows the damped oscillation of the laser field after switch-on (relaxation oscillations). As the oscillation is damped, the laser settles for longer times to a cw emission corresponding to a time-independent solution of the laser equations that is stable, which means that small perturbations of the laser will be damped out. Thus cw laser emission is also called stable laser emission.

The solution of the equations may also be *unstable*, which means that small perturbations of the laser will be amplified. The laser can then attain another emission form, e.g. corresponding to a time-periodic solution reflected in regular periodic pulsing

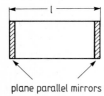

plane parallel mirrors

Fig. 1.1.2. Linear resonator with plane mirrors. The mirror distance is l.

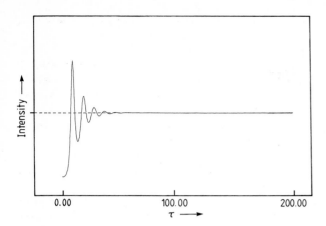

Fig. 1.1.3. Damped relaxation oscillations after switching on a laser. The asymptotic value of the intensity corresponds to the stationary (cw) laser emission.

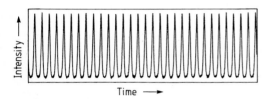

Fig. 1.1.4. Regular pulses emitted from a laser.

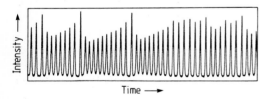

Fig. 1.1.5. Non-periodic, irregular pulses emitted from a laser ("chaotic" pulsing).

of the laser output (Fig. 1.1.4). The time dependence (pulsing) of the laser output power can be accompanied by a corresponding time dependence of the laser frequency.

Another emission form that is possible when the time-independent solution is unstable is non-periodic pulsing, also termed "chaotic" pulsing. In this case, although there is still some regularity or periodicity in the output, the exact form of pulsing never repeats and many characteristics of the laser output show stochastic properties (Fig. 1.1.5). The spectrum of the laser field or intensity, for example, is continuous and noise-like. Evidently with the intensity the laser frequency may vary correspondingly non-periodically.

It is also possible that time-dependent behavior corresponds to an unstable time-dependent solution. In this case it is possible that, e.g., the laser emits for a long period in a chaotic manner before it can finally settle to cw or periodic emission ("metastable chaos") (Fig. 1.1.6).

The typical pulsing periods are usually associated with the relaxation oscillation mentioned above. As the cw emission gradually becomes less stable, the damping of the relaxation oscillations weakens. At the point where the cw emission has become unstable, the relaxation oscillations become undamped. This laser pulsing is therefore also termed "sustained relaxation oscillations".

Multiperiodic pulsing is often also encountered. Figure 1.1.7 shows so-called "period-4" pulsing. Apart from these dynamics pertaining to emission of only one laser frequency, a wide variety of time-dependent emission forms, so far largely unexplored, are encountered when the laser emits several frequencies simultaneously (multimode

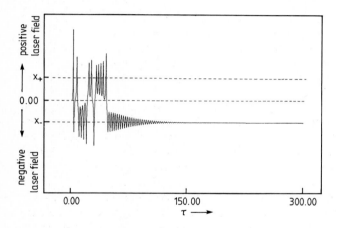

Fig. 1.1.6. "Metastable chaos". The laser shows a long transient chaotic pulsing before settling to its final cw emission. In this figure, the field strength, which can be either positive or negative for the individual pulses, is shown instead of intensity. This is a good example of the behavior of the simplest kind of laser.

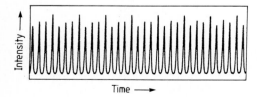

Fig. 1.1.7. "Period-4" pulsing of a laser as an example of multiperiodic laser pulsing.

emission) and the radiations of the different frequencies are allowed to influence each other. The best known and simplest case is that of "longitudinal mode-locking". In this case, the radiation frequencies of the laser synchronize in phase.

As a result of interference, very short radiation pulses are then created. This ultrashort pulse generation is a vast field in itself; however, in general, the interaction of several frequencies emitted by one laser creates dynamics, chaotic or regular, often not only with a time dependence but also differing for different points in space in the laser.

1.2   Stable Lasers and Laser Instabilities

Until the early 1980s, the concept of laser operation was largely associated with stable laser emission (stable in the mathematical sense that smooth changes of the laser parameters bring about smooth changes in the laser output power or the laser frequency). Hence constant pumping should yield cw laser emission and pulsed pumping should give rise to a radiated pulse, the shape of which reflects the time dependence of the pump. Specifically, a single-mode laser was assumed always to emit stably and a homogeneously broadened multimode laser was assumed to develop into single-mode stable emission by mode competition. For inhomogeneous broadening, the case of independent oscillations of several modes was usually considered, this again leading to a stable laser output.

Of course, it was well known that pulsing lasers exist; that in particular these in some cases show large, not completely understood and uncontrollable pulse-to-pulse power variations; that solid-state lasers tend to emit in irregular bursts of short pulses rather than in a smooth pulse reflecting the shape of the time-dependent pump; that multimode lasers may show either irregular time-dependent outputs (Fig. 1.2.1) or trains of equally spaced short pulses ("mode-locking"); and that lasers with saturable absorbers can generate repetitive short pulses under constant pumping ("Q-switching"). However, the ununderstood cases were explained by relatively simple arguments: insufficient control over the laser parameters (such as pump power or resonator length) which give rise to gain fluctuations, insufficient time for mode competition to develop into a single-mode state, spatial hole-burning or even insufficient experimental skill.

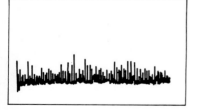

Fig. 1.2.1. The continuous high-frequency relaxation oscillations in the ruby-maser output. The recorded time, is 500 μs. From Nelson and Boyle, 1962.

The mode-locked and passive Q-switching regimes were perceived as artificial exceptions of the generally stable laser behavior. Accordingly, little effort was spent in trying to understand the phenomena, e.g. to explore the limits of the passive Q-switching regime, or in trying to understand the "instabilities" of the mode-locked operation which limit the attainable shortest pulse lengths.

Even though the idea of the laser as a "stable" device prevailed, the contrary has been known, at least theoretically, for almost as long as lasers have been known. For example, in 1964, Grazyuk and Oraevskii found in numerical studies of the equations describing the simplest laser (a homogeneously broadened, single-mode, traveling-wave, resonantly tuned laser), a time-dependent solution that consisted of pulses, varying irregularly with time (Grazyuk et al., 1964). They even used at that time the term "chaotic" to describe the irregular pulsing behavior they found.

H. Haken (1975) recognized that the laser equations are isomorphic with those used by Lorenz (1963) to study Rayleigh-Bénard convection flow. In order to make this problem tractable, Lorenz decomposed the flow variables into Fourier series and retained only the first terms of the series. The resulting three ordinary first-order differential equations exhibited non-periodic flow solutions. Lorenz appears to have realized the importance of this result: simple equations can have very complex solutions, which gave indications along which lines the phenomenon of fluid turbulence might possibly be understood. From the isomorphy of the laser with the Lorenz equations, Haken concluded that lasers should exhibit non-periodic, pulsing emission: "chaotic" emission. In the laser (or Lorenz) equations, the chaotic solution appears abruptly with large amplitude pulses above a certain pump strength in the manner of a mathematical instability, smooth variations leading to a sudden, abrupt change. This pump strength was named the "second laser threshold", in analogy with the ordinary (first) laser threshold, at which laser emission suddenly sets in in a manner analogous to a second-order phase transition.

The identification of the second laser threshold, associated with the onset of chaotic (turbulent) emission, did not really change most experimentalists' view of the laser as a "stable" device because:

1) the conditions for which this second threshold is reached requires a laser resonator loss that is unrealistically high for common lasers; and

2) as it already seemed difficult to achieve laser emission in such a lossy resonator, it seemed even more unrealistic to reach the second laser threshold, which occurs at $10-20$ times the pumping necessary to reach the first laser threshold.

The Lorenz-Haken instability therefore appeared to be only an academic curiosity invented by theoreticians far remote from the day-to-day reality of experimental laser physics. Theoreticians shared this view. The Lorenz model was studied largely because it constitutes a beautiful, rich, but still simple case of chaotic dynamics. From its study, much basic understanding of chaotic system behavior could be drawn. The Lorenz equations in fact to this day play the role of the "guinea pig" in theoretical studies of non-linear dynamics.

Interest in laser instabilities, e.g. the spontaneous onset of pulsing, arose in the early 1980s as a consequence of several new developments, as follows.

1) The work of Casperson (1978) on the high-gain Xe-laser had revealed that the second laser threshold (spontaneous onset of periodic pulsing) of inhomogeneously broadened lasers is substantially lower than that of the homogeneously broadened laser.

2) The mathematical work of Ruelle and Takens (1971) had shown that chaotic solutions of non-linear differential equations can occur after passing through only a small number of instabilities. This was in contrast with the Landau picture of turbulence, which postulated that turbulence would occur after an infinite number of instabilities, each exciting a new mode with a new frequency, so that the motion, governed by an infinite number of frequencies, would be very complex.

3) Feigenbaum (1978) found a particularly simple example of the onset of chaos in which each new frequency occuring at an instability point is simply one half of the frequency excited at the previous instability point. He was able to prove that this type of onset of chaos is typical of a large class of non-linear processes.

4) Convincing experimental evidence for the onset of chaos according to the Feigenbaum and Ruelle-Takens pictures had been found in hydrodynamic systems, in particular in Bénard convection flow experiments (Maurer and Libchaber, 1979; Libchaber and Maurer, 1980).

5) Clear evidence of the Feigenbaum onset of chaos had also been found in a so-called hybrid electro-optical device, an arrangement meant to model passive optical resonators containing a non-linear medium (H.M. Gibbs et al., 1981).

6) Theoretical suggestions had been made about the way in which the number of degrees of freedom of typical laser systems might be enlarged in order to generate a chaotic dynamic evolution of the laser output. This could be accomplished by introducing an additional optical field, by modulating parameters such as the population inversion or by admitting several modes (Yamada and Graham, 1980; Scholz et al., 1981).

The more widespread notion of the laser as a system that potentially exhibits instrinsic dynamics and instabilities resulted in two experiments giving the first clear evidence of chaotic dynamics of simple lasers: observations on a loss-modulated CO_2-laser (Arecchi et al., 1982) and on a multimode infrared He$-$Ne laser (Weiss et al., 1983). These results, sharpening the perception of lasers as "unstable" systems, were then followed by a large number of experimental and theoretical investigations, which were motivated by the following (among others):

1) The practical question of finding where the limits for stable laser emission are. The answer to this question is important for high-power laser operation and for lasers to be used in optical communication systems, but more generally for all technical applications of lasers.

2) Possible applications of lasers with a well defined incoherence.

3) A better understanding and control over lasers producing ultrashort pulses.

4) The interest in the dynamics of non-linear systems and non-linear optical systems in general.

Generally, lasers allow experiments to be performed concerning widely different fields of non-linear dynamics:

In fluid dynamics terminology, a single-mode laser is a "highly confined" system. This means that almost all of its properties (such as spatial structure and the optical spectrum of the laser field) are rigidly determined by the boundary conditions (mirrors, apertures). Thus, for a single-mode laser hardly anything other than the temporal behavior can be seriously influenced by the non-linearities.

In contrast, a"weakly confined" system is largely governed by its non-linearities and its time constants and is very little influenced by boundary conditions. Evidently, weakening boundary conditions is possible in lasers, and the simplest consequence is a multimode emission. The interaction of these modes, mutually and with the non-linear medium, will then produce very rich spatio-temporal dynamics more resembling the better known instances in non-linear dynamics, such as fully developed turbulence or all kinds of spontaneous pattern formations or pattern dynamics.

The term "instabilities" so far refers to temporal properties such as the onset of periodically or chaotically pulsing laser output, and research has been concerned almost exclusively with these. Recently, however, the first theoretical suggestions of spatial instabilities in lasers and optics, i.e. the spontaneous formation or change of spatial structures, have appeared. These may be spatially periodic or chaotic. Combined with temporal instabilities, one may expect them to lead to the first experimentally well controllable (as they are sufficiently simple) systems displaying genuine turbulence, i.e. chaos in space and time.

1.3 Lasers and their Relation with Instabilities in Dissipative Systems

Instabilities and chaotic motion occur in both conservative and dissipative systems. Whereas a conservative system never loses the memory of its initial conditions, dissipative systems have asymptotic solutions for $t \rightarrow \infty$ that can be independent of the initial conditions. This makes their treatment easier and simplifies their phenomenology. Lasers are dissipative systems: on the one hand energy is supplied to them by pumping, and on the other energy is dissipated via the outcoupling of light and also relaxation processes, which include conversion into heat within the active medium. If the transients (i.e. the initial period required for the laser to forget its initial conditions, for instance after switching on or after crossing an instablility point) are ignored, the laser can be described by the asymptotic solutions ("attractors") (see Chapter 3) of its differential equations.

Fluid flows: Lasers bear a resemblance to fluid flows. As mentioned in the previous section, the equations describing simple lasers are isomorphic with a truncated description of convective flow in cells. Energy can be supplied to a rectangular cell, containing a fluid, by a temperature difference between its bottom and upper sides. With a small difference one observes heat conduction. In the laser case, this corresponds to pumping with a strength insufficient to reach the laser threshold. In the cell no

macroscopic order occurs; the laser rids itself of the pump energy by radiating incoherent, unordered spontaneous emission.

With a larger temperature difference, circular macroscopically ordered flow patterns,
or "rolls", form spontaneously in the cell as a consequence of the transition from heat
conduction to heat convection. In the laser this corresponds to pumping above the
threshold, where the spontaneous formation of the spatially and temporally ordered
(coherent) emission mode has taken place. The circular, time-periodic motion in the
convection cell may be thought of as the counterpart of the oscillation at the optical
laser frequency.

With an even larger temperature difference, the convection cell becomes turbulent
as the laser at high pump strength becomes chaotic.

The equations describing in a simplified way Bénard convection and, as will be
described in Chapter 2, the simplest form of a laser are

$$\dot{x} = \sigma(y - x)$$
$$\dot{y} = x(r - z) - y \qquad\qquad (1.3.1)$$
$$\dot{z} = xy - bz$$

For the laser case x represents the electric field, y the polarization and z the population inversion. The meanings of the parameters b, r and σ for the two cases are as
follows:

Bénard convection	Laser
σ: Prandtl number	$\sigma = \kappa/\gamma_\perp$
$r = R/R_C$	$r = \lambda + 1$
$b = \dfrac{4\pi^2}{\pi^2 + k_1^2}$	$b = \gamma_\parallel/\gamma_\perp$

where r is the Rayleigh number R normalized to the value that it takes at the onset
of convection, k_1 is a dimensionless wavenumber, κ represents the damping rate of the
laser resonator, λ is the pump rate above the threshold ($\lambda = 0$ at the laser threshold),
$\gamma_\parallel$ is the energy relaxation rate in the laser medium and $\gamma_\perp$ is the polarization relaxation
rate. For $\lambda = 0$ an instablility is found that corresponds to onset of convection ($r = 1$) and continuous laser emission.

Thus, the study of the solutions of (1.3.1), which will be given in detail in Chapter
2 and 3, applies not only to lasers but also to fluid flows. Let us only mention here
the fact that, if certain conditions are met, a threshold for the onset of instabilities is
found. Therefore, two kinds of thresholds can be defined for a laser: the "first laser
threshold", which corresponds to the well known threshold for the stable laser emission
and is equivalent to a second-order phase transition, and the "second laser threshold",
which gives rise to the onset of unstable behavior.

Population growths: The Lorenz equations display a particularly simple onset of
chaos after only one instability. In contrast, as mentioned in Section 1.2, other simple

systems require crossing of an infinite number of instability points before chaotic motion sets in.

For the study of population growth, in 1845 Verhulst introduced the following model. In a limited area of land, the number of species x_{n+1} in the year $n+1$ is proportional to the number x_n existing in the preceding year and to the area remaining unoccupied by the population x_n. Thus,

$$x_{n+1} = r(1 - x_n)x_n$$

where r depends on the fertility, the supply of food from the area, etc.

The population x does not necessarily stabilize in the long term. Depending on r, it can either, reach a steady state, or be periodic with a simple or complicated structure of the periods, or behave chaotically. It is found that as r is increased, the population initially stabilizes, and at larger r it becomes periodic. As r is increased further an infinite number of values of r_n are encountered at which the period lenghts double from the preceding periods. These instability points converge towards an accumulation point r_∞, at which the period, obviously, is infinite. In other words, the x_n form a non-periodic time series. See Fig. 2.2.1. Feigenbaum (1978) was able to show that this behavior is not limited to the population growth equation but occurs for a wide range of equations.

The laser is again related. If the laser equations are extended to allow for detuning of the laser resonator with respect to the medium gain line center, one finds predominantly transtions to chaos via period doubling, as in the Verhulst equation, when the pump strength or resonator tuning is varied (Zeghlache and Mandel, 1985).

Driven non-linear pendulum (and electrical equivalent): One of the simplest systems showing a period-doubling transition to chaos is the driven pendulum or the electrical equivalent, the non-linear, periodically driven resonator. Among a wide variety of driven non-linear resonators, all of which display period-doubling, is a special case of lasers. If the polarization of the laser medium is of no importance, the laser is a system in which energy can be stored in two ways: in the form of population inversion of the medium or in the optical field. The energy can oscillate back and forth between these two reservoirs, giving rise to the well known damped relaxation oscillations of lasers (Stats and de Mars, 1960). Periodic modulation of one laser parameter at a frequency related to the relaxation oscillation resonance frequency then gives rise to period doubling and chaotic laser emission. Chaotic laser dynamics were achieved for the first time in this way (Arecchi et al., 1982).

The dynamics of lasers of even the simplest type are thus closely related to basic models of non-linear dynamics and hence to a variety of phenomena observed in other areas of physics and in other disciplines of science. The weather and economic, social, chemical and biological systems nowadays appear to be governed by chaotic dynamics. In contrast to these, the laser is a relatively simple system and, unlike the large majority of them, it is possible with lasers to control their degrees of freedom. Studies of laser dynamics may therefore be expected to provide increasing insight into the dynamics of non-linear systems in general, by gradually progressing from the most "confined" to less and less "confined" cases. The results of the study of laser dynamics might therefore, apart from their practical interest, prove useful for understanding complex penomena that have so far not been understood. A good example is the case of multimode dye lasers described in Chapter 9. The temporal behavior of individual modes in this case is found to be irregular. The number of degrees of freedom of this laser is enormous. However, a determination of fractal dimension shows that only three variables govern the systems: the inversion and two optical fields (the reason why two fields are relevant originates from the "spatial hole burning"). Thus one might be optimistic that complex phenomena such as those encountered in everyday life may find simple descriptions in terms of low-dimensional chaotic dynamics.

1.4 Properties of Lasers

Lasers are usually classified according to the material that provides the optical amplification. This material determines largely the properties of the laser: the mode of operation (pulsed or continuous), the emission wavelength, the output power/energy and the coherence properties. Gases, liquids and solids can provide optical amplification when properly excited.

Solid-state lasers are usually excited optically by pulsed discharge lamps for high-power pulses and by continuous lamps or lasers (or the sun) for continuous operation. The exception is the semiconductor lasers, which can be excited optically, by electron beams or by passing current through a $p-n$ junction (diode lasers).

Laser liquids are usually dye solutions which may be excited by lamps or lasers for pulsed operation or by continuous lasers for continuous operation.

Gas lasers are usually excited by a high-voltage glow discharge. There are other types which use chemical excitation or optical excitation by lamps and lasers or even by combustion.

The laser transition of the amplifying material may be homogeneously broadened, i.e. light of a certain optical frequency can interact with all atoms/molecules, all of them having the same resonance frequency. The homogeneous linewidth, $\Delta \nu_{\mathrm{H}}$, is given by the medium relaxation rates:

$$\gamma_{\perp} + \gamma_{\parallel} = \pi \Delta \nu_{\mathrm{H}}$$

where $\gamma_{\parallel}$ and $\gamma_{\perp}$ are the relaxation rates for inversion and polarization, respectively. In the inhomogeneous broadening case, the material consists of atoms/molecules of different resonance frequencies. Light of a particular optical frequency can then only interact with a fraction of the total number of atoms/molecules. The different resonance frequencies can be caused in gases by the Doppler effect of the moving atoms or in solids by local electric fields, Stark shifting the atomic transition by different amounts.

As regards laser dynamics, lasers may be classified in still another way. Lasers operating in a single emission mode are described by the three equations (1.3.1). The three relevant variables, field, population and polarization, usually decay on very different time scales, which are given by the relaxation rates κ, $\gamma_\parallel$ and $\gamma_\perp$. If one of these is large compared with the others, the corresponding variable relaxes fast and consequently adiabatically adjusts to the other variables (see Chapter 3). The number of equations describing the laser is then reduced.

Those lasers for which the population and the polarization decay fast in comparison with the field have been named class A lasers, those for which only the polarization relaxes fast, class B lasers, and those for which all three relaxation rates are of the same magnitude, class C lasers. Table 1.4.1 lists, as examples, lasers in the various categories.

The laser equations for a class A laser reduce to one. Therefore, only constant output solutions exist. For the class B lasers, oscillation of the energy between field and inversion is possible and the equations yield relaxation oscillations. Class C lasers with their coupled dynamics of field, inversion and polarization can display undamped periodic or non-periodic (chaotic) pulsing.

Class B lasers can, however, show chaotic dynamics when they are externally influenced (modulation of a parameter, injection of external light, feedback). As for the time-dependent phenomena, possible for class C single-mode lasers, the medium polarization of the atomic coherence plays a significant role. These phenomena therefore have to be counted among the "coherent effects" in the interaction of radiation with matter, such as the photon echo (Brewer, 1977), free induction decay (Brewer and Shoemaker, 1972) or the optical rapid adiabatic passage (Treacy and De Maria, 1969).

Table 1.4.1. Laser classes.

Active material	Pumping	Broadening	Class
CO_2	discharge, optical, laser	homogeneous	B
Ar^+, Kr^+	discharge	inhomogeneous	C
He-Ne	discharge	inhomogeneous	C
He-Xe	discharge	inhomogeneous	C
He-Cd	discharge	inhomogeneous	C
Ruby	lamp, laser	homogeneous	B
Nd:YAG	lamp, laser	homogeneous	B
Nd:glass	lamp, laser	inhomogeneous	B
Alexandrite	lamp, laser	homogeneous	B
GSGG	lamp, laser	homogeneous	B
Ti:sapphire	lamp, laser	homogeneous	B
$Ni,CO:MgF_2$	lamp, laser	homogeneous	B
color-center	lamp, laser	homogeneous	A
Dye	laser	homogeneous	A
Semiconductor	diode, laser	homogeneous	A
Optically pumped gas laser	laser	homogeneous*	C
Chemical gas laser (HF; I)	chemical reaction	homogeneous	B

* Sometimes with inhomogeneous effects due to coherent pumping.

All of these are distinguished from other interactions by the relevance of the phase of the atomic wavefunction, which manifests itself in interference phenomena with the optical field.

From the non-linear dynamics viewpoint, experiments with lasers offer advantages over similar investigations with fluids. Since the basic oscillation is an optical frequency, all characteristic frequencies are very high. Hence the requirements for parameter stability for measuring the time-dependent behavior for the long periods necessary for quantitative investigations are easier to fulfill than in fluid experiments. Pulsing frequencies are typically 6–8 orders of magnitude higher than for fluid systems, with a corresponding reduction in the observation time for lasers.

The number of degrees of freedom of lasers, e.g. the number of modes, can be easily controlled. This is again different from fluids, where boundary layers present their own difficult-to-control dynamics. In laser experiments, the relevant parameters are usually well known and controllable and the experimental conditions are well defined.

2 Some Basic Illustrative Examples

The typical features of non-linear dynamic behavior can be better understood if we begin by describing in detail some simple illustrative examples. In this chapter, the Lorenz-laser model and the "logistic map" have been chosen as examples.

2.1 The Lorenz-Haken Model

2.1.1 The Lorenz Equations

As was pointed out in the preceding chapter, Haken (1975) realized that the equations ruling the simplest laser, i.e. a homogeneously broadened single-mode traveling-wave resonantly tuned laser, are isomorphic with the equations obtained by Lorenz (1963) for the description of convection flows in fluids. These simple equations are usually expressed in the following form:

$$
\begin{aligned}
\mathrm{d}x/\mathrm{d}\tau &= -\sigma(x - y) \\
\mathrm{d}y/\mathrm{d}\tau &= rx - y - xz \\
\mathrm{d}z/\mathrm{d}\tau &= xy - bz
\end{aligned}
\tag{2.1.1}
$$

where the variables x, y and z are real, τ is proportional to time and the parameters σ, r and b are real and positive. As is shown in Chapter 5, in the case of lasers the meaning of the variables and parameters is as follows:

$$x = \text{field amplitude}$$
$$y = \text{induced polarization}$$
$$(r - z) = \text{population inversion} \qquad (2.1.2)$$
$$\tau = \gamma_\perp t$$
$$\sigma = \kappa/\gamma_\perp$$
$$b = \gamma_\parallel/\gamma_\perp$$
$$r = D_o/\bar{D}$$

where κ, $\gamma_\perp$ and $\gamma_\parallel$ represent the relaxation rates for the field amplitude, polarization and population inversion, respectively. D_o and $\bar{D}$ symbolize the stationary population inversions in the absence or presence, respectively, of a generated field. Note that r gives a measure of the pumping mechanism strength and $z = r - D/\bar{D} = (D_o - D)/\bar{D}$, where D represents the instantaneous population inversion, so that the stationary value of z in the presence of a field is $r - 1$.

The system (2.1.1) is composed of three coupled first-order equations that contain only two non-linear terms of low order ($-xz$ and xy). In spite of their simplicity, the presence of these terms gives rise to a very rich and unexpected dynamic behavior, which is described below. To this end, let us consider a given laser system with fixed values for the relaxation parameters σ and b, and let us increase the pumping rate. This simply means that r is increased from zero to a large value.

As we do this, we find two kinds of solutions of eqns. (2.1.1): "normal" ones which appear for low r, and "strange" ones which appear for large r. Both of them are described below.

2.1.2 "Normal" (Stable) Solutions

When r is increased from zero, different domains can be distinguished, as follows.

a) $0 \leqslant r \leqslant 1$.

In this domain the only stationary solution is the zero point: $(x, y, z) = (0, 0, 0)$. Whatever the initial conditions may be, the "trajectory" of the system in the phase space defined by the variables x, y and z eventually approaches the zero point. This simply means that the pumping rate is too low for the gain to exceed the losses and there is no laser emission.

b) $1 \leqslant r \leqslant r_A$.

For $r > 1$ the zero point solution still exists, but it is now unstable. Simultaneously, two other stationary stable solutions, C_+ and C_-, appear:

$$C_\pm = (\pm\sqrt{b(r - 1)}, \pm\sqrt{b(r - 1)}, r - 1). \qquad (2.1.3)$$

Physically, these two solutions are almost identical, because they only differ in the sign of x and y, i.e. in the phase of the field and polarization. In fact, eqns. (2.1.1) are symmetric with respect to the change $(x, y, z) \rightarrow (-x, -y, z)$, so that for every solution $(x(t), y(t), z(t))$ the symmetric solution $(-x(t), -y(t), z(t))$ also exists. The points C_+ or C_- represent the stationary state yielding cw laser emission when the gain is larger than the losses.

These results mean that if we start with the laser system near the zero point in the phase space, the trajectory moves away from it and eventually approaches the C_+ (or C_-) point and then always remains on it. This is illustrated by the numerically calculated examples in Figs. 2.1.1–2.1.3,, for which typical values corresponding to a specific type of laser (an ammonia far-infrared laser; see chapters 5 and 8) have been assigned to the σ and b parameters. Independent of the exact values of the initial conditions, any trajectory initiating close to the zero point moves away from it in a fixed direction (either in the positive or negative sense); however, for $r < 4.9$ (the case in Fig. 2.1.1) the trajectories tend towards the point C_+; for $4.9 < r \leqslant 16.713$ (the case in Fig. 2.1.2), the trajectories first turn once around C_+ but then spiral around C_-; and for larger values of r up to $r_A = 35.850$ (the case in Fig. 2.1.3), the trajectories first spiral in an apparently random fashion around C_+ and C_- and finally they are eventually attracted to one of these points. The transient random spiralling around C_+ and C_- is known as "preturbulence" or "metastable chaos".

The change from Fig. 2.1.1 to Fig. 2.1.2 occurs when the trajectory, after a turn around C_+, eventually approaches (returns to) the zero point (dashed line in Fig. 2.1.2.a). In a similar way, the change from Fig. 2.1.2 to Fig. 2.1.3 occurs when the trajectory, after a turn around C_+ and a turn around C_-, eventually returns to the zero point. This kind of orbit which returns to the initial point is known as a "homoclinic orbit".

Even if in the three cases in Figs. 2.1.1–2.1.3 the asymptotic behaviour (i.e. for long times) is qualitatively the same, it is evident that the trajectories in the phase space (let us think of all the possible trajectories that could be traced in each case, which constitute the "phase portrait") have experienced qualitative changes at some specific values of the pumping rate ($r = 1, 4.9, 8, \ldots, r_A$). These "structural changes" appearing when a parameter is varied are called "bifurcations", which are analysed and classified in the next chapter.

2.1.3 "Strange" Solutions

If we further increase the pumping rate r beyond r_A, we find several coexisting kinds of solutions, as follows.

a) *Strange attractor*

When r exceeds the value $r_A = 35.850$, a trajectory initiating near the origin displays a "strange" behavior: instead of asymptotically approaching the points C_+ or C_-, as

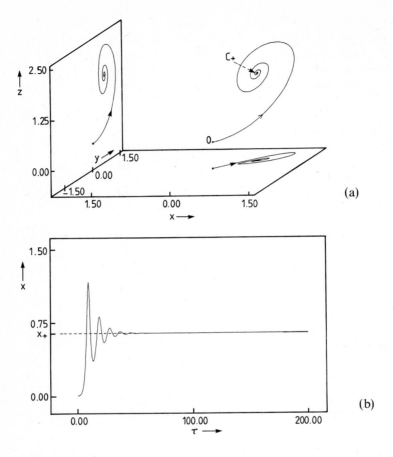

(a)

(b)

Fig. 2.1.1. (a) A trajectory in the phase space originating close to the zero point and eventually approaching the point C_+, which represents the stationary state yielding cw laser emission. (σ = 1.4253, b = 0.2778, r = 2.5). (b) The x-projection of the same trajectory as a function of time (x_+ is the x-component of C_+).

in the previous figures, it now remains forever spiralling around these points (Fig. 2.1.4): after a random number of turns around C_+ it changes to C_-, returns to C_+, and so on. There is no periodicity: the trajectory remains in the zones around C_+ and C_- but never closes on (or intersects) itself. The intensity of the generated field behaves as an irregular or "chaotic" sequence of spikes, Fig. 2.1.4.c.

The zone of the phase space where the trajectory remains (Fig. 2.1.4.a) is called an "attractor", because it attracts not only the trajectories initiating close to the origin but almost all the trajectories originating outside of it. The attractor does not look like a thick (i.e. three-dimensional) body, but rather like a curved sheet. In fact, there is something surprising about the spatial dimensions of the attractor: on the one hand, its volume must be zero (like a line or a surface). Since the Lorenz equations (2.1.1)

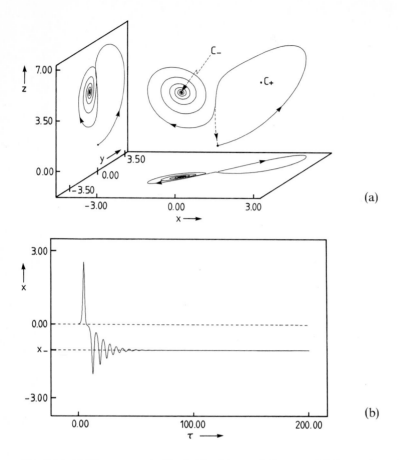

(a)

(b)

Fig. 2.1.2. (a) The same as in Fig. 2.1.1, with $r = 5$. For $r \approx 4.9$ there exists a "homoclinic orbit" (dashed line). (b) The x-projection of the same trajectory as a function of time (x_- is the x-component of C_-).

contract volumes in the phase space as a result of the presence of relaxation terms: any volume V evolves with time as

$$V(t) = Ve^{-(\sigma + b + 1)t}. \tag{2.1.4}$$

On the other hand, however, the attractor must have a dimension larger than two, in order to prevent a trajectory from intersecting itself (intersection is forbidden for solutions of first-order differential equations); this point is illustrated in the central part of Fig. 2.1.4.a, or in Fig. 2.1.5, which represents the intersection points of the trajectory with the plane $z = r - 1 = 39$ in this central part ("Poincaré map").

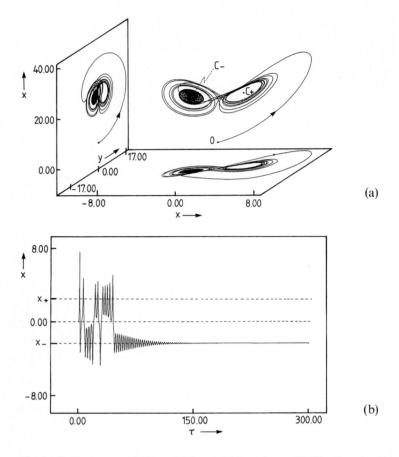

(a)

(b)

Fig. 2.1.3. The same as in Figs. 2.1.1 and 2.1.2, with $r = 20$. The "metastable chaos" is shown in (b).

The only solution to this apparent contradiction is that the attractor's dimension be between 2 and 3, i.e. it should be "fractal". In fact, the dimension has been calculated and its value is 2.06 (Hausdorff dimension, which is defined in Section 4.2.2).

An approach to understanding how the dimension of an attractor in a three-dimensional phase space can be larger than two consists in following the time evolution of an initial cubic volume: it diminishes, but it does not mean that it contracts in all the spatial directions; in one or two directions it continuously stretches and folds over on itself an unlimited number of times (like dough being kneaded) (Sparrow, 1982), so that finally, and loosely speaking, it looks like a sheet with some kind or small degree of "effective thickness", i.e. with a dimension larger than two but lower than three.

At present there is extraordinary interest in "fractal objects" in mathematics and science. Known examples are those of Fig. 2.1.6, which are intermediate objects between

(a)

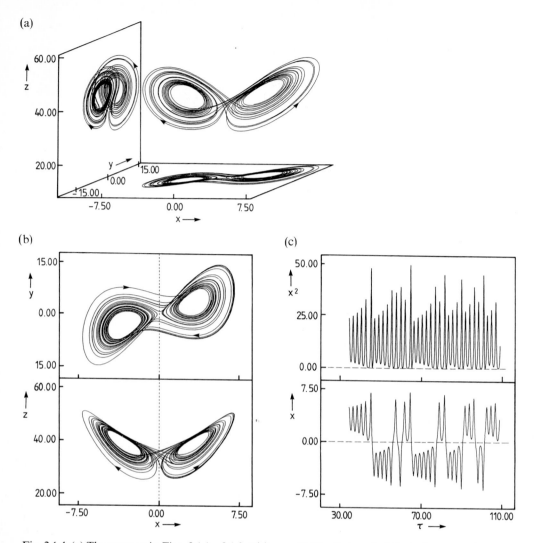

Fig. 2.1.4. (a) The same as in Figs. 2.1.1–2.1.3, with $r = 40$. The first part of the trajectory, going from the neighborhood of the zero point to the zone of spiralling around C_+ and C_- (transient regime), has not been plotted (the points C_+ and C_- lie in the "holes" of the attractor). (b) $x - y$ (upper curve) and $x - z$ (lower curve) projections of the trajectory in (a). (c) Lower curve: the x-projection (i.e. the laser field) of the trajectory plotted in (a) as a function of time. Upper curve: the square of the x-projection (i.e. the laser intensity) as a function of time.

a set of points and a line (dimension between 0 and 1, (a)) or between a line and a band (dimension between 1 and 2, (b)). Note that the Koch curve (b) has an infinite length buth encircles a finite area.

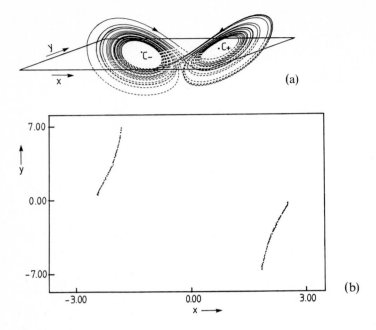

Fig. 2.1.5. (a) The same attractor as in Fig. 2.1.4.a. The plane $z = r - 1 = 39$ has also been plotted. (b) Intersection points of the trajectory plotted in (a), with the plane $z = r - 1 = 39$. Only the points created when the trajectory crosses the plane downwards have been considered. The sequence of these points, taken in order of appearance, defines a "Poincaré map".

The attractor in Fig. 2.1.4. has another striking property, also due to the stretching of lengths in some directions, i.e. a "sensitive dependence on the initial conditions". This means that two trajectories originating at close together points will eventually diverge (Fig. 2.1.7). At the beginning the separation increases exponentially and, when it reaches a value of the order of the linear dimensions of the attractor, the trajectories remain within it but are uncorrelated. The time necessary to reach this final situation is often not very long. This has the important physical consequence that it is almost impossible to make long-term predictions, because in practice it would require being able to reproduce initial conditions with an extremely high precision. Dynamic behaviors (or attractors) with a sensitive dependence on initial conditions are called "chaotic".

Usually fractal dimension (a geometrical property) and chaotic character (a dynamical property) are properties that are found together. An attractor showing these properties is known as a "strange" attractor.

b) *Fixed points*

Let us consider the "fixed points" C_+ and C_-, which, for $1 < r < r_A$, were attracting any trajectory (Figs. 2.1.1 – 2.1.3). For $r > r_A$ they still exist but now they only attract

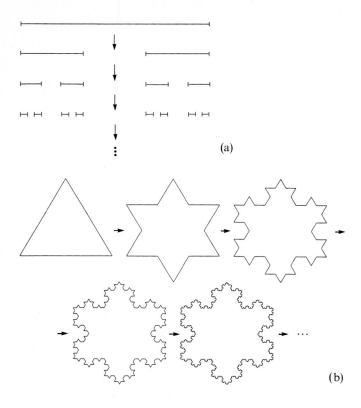

Fig. 2.1.6. Fractal objects obtained through an infinite sequence of steps: (a) a Cantor set, obtained by dividing each segment into three parts and taking out the central one (Hausdorff dimension = 0.6309); (b) Koch curve, obtained by addition of a new triangle to each side of the previous figure (Hausdorff dimension = 1.2618).

the trajectories originating close to them, i.e. in the "holes" found inside the strange attractor (Fig. 2.1.4-a). With increasing r the size of these holes or "basins of attraction" diminishes, and at

$$r = r_H = \sigma(\sigma + b + 3)/(\sigma - b - 1) \qquad (2.1.5)$$

($= 45.45$ in the example considered here) they disappear. This means that the stationary solutions C_+ and C_- still exist, but they are unstable for $r_H < r < \infty$ (like a ball on top of a hill). Thus, over a wide range of values of r going from $r = r_A$ to $r = r_B$ ($r_B \approx 138.5$ in our case), the laser dynamics is mainly determined by the properties of the strange Lorenz attractor. Note that a necessary condition for the fixed points C_+ and C_- to become unstable (i.e. $r_H < \infty$) is that

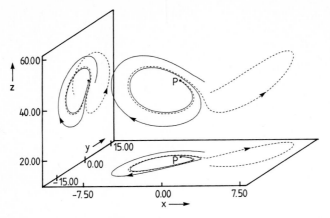

Fig. 2.1.7. Two trajectories (continuous and dashed lines, respectively) originating at close together points (near the point P) and rapidly diverging. They become completely uncorrelated within the attractor.

$$\sigma > b + 1 \tag{2.1.6}$$

which in the case of a laser (see (2.1.2)) means

$$\kappa > \gamma_{\parallel} + \gamma_{\perp}. \tag{2.1.7}$$

This inequality is known as the "bad cavity condition", because it requires that the cavity losses for the field be larger than the sum of the population inversion and polarization relaxation rates. The values for the pumping parameter $r = 1$ (i.e. onset of laser emission) and $r = r_{\mathrm{H}}$ (i.e. onset of unstable emission) determine the so-called "first" and "second laser thresholds", respectively, which are equivalent to phase transitions.

c) *Periodic "windows"*

The Lorenz attractor is not strange for the whole domain $r_{\mathrm{A}} < r < r_{\mathrm{B}}$. In fact, there exist small intervals, called "periodic windows", in which the attractor is periodic. For instance, in the considered case two periodic windows are found, for

$$107.3 < r < 117.8$$
$$73.4 < r < 73.5$$

Figure 2.1.8 shows an example of this periodic behavior. Other windows exist, but most of them are so narrow that they are physically irrelevant (the value of r and of other parameters is affected by experimental noise and fluctuations).

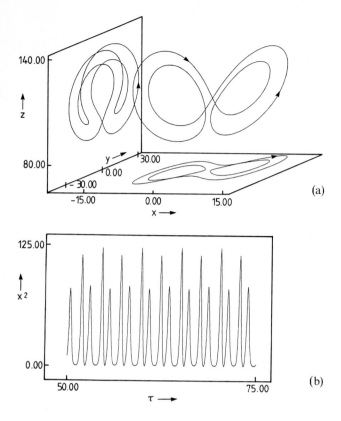

(a)

(b)

Fig. 2.1.8. (a) A periodic trajectory in the phase space, for $r = 110$. It corresponds to a "periodic window". (b) Time evolution of the variable x^2 (field intensity).

2.1.4   A "Road to Chaos"

Let us analyse the last domain of r values, $r_B \approx 138 < r < \infty$, which is characterized by the fact that the attractor loses its strange character and becomes periodic through a sequence of bifurcations. As this sequence is more easily described in the opposite order (i.e. going from periodic to chaotic behavior), let us first consider the case of very large r and then progressively decrease it to r_B.

For $r_C \approx 146.5 < r < \infty$ the attractor is a simple closed curve (Fig. 2.1.9.a). This represents a periodic pulsing for the field (Fig. 2.1.9.b), whose period T changes very slowly with r.

At $r = r_C$ the curve experiences a qualitative change: it closes on itself not after one but after two round trips (Fig. 2.1.10), so that the period changes from T to $2T$. This is known as a "period-doubling" bifurcation.

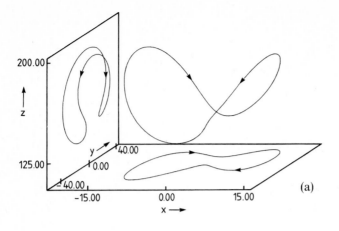

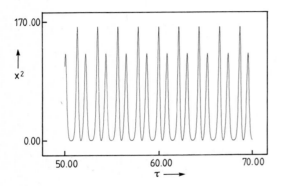

Fig. 2.1.9. (a) Periodic attractor corresponding to $r = 160$ (other parameter values as in Fig. 2.1.1.a); (b) time dependence of the field intensity.

This shape for the attractor remains for $r_{C'} \approx 141.5 < r < r_C$. At $r = r_{C'}$ a new "period-doubling" bifurcation appears: the curve closes on itself only after four round trips (Fig. 2.1.11), so that the period suddenly changes to $4\,T$. This structure holds for $r_{C''} < r < r_{C'}$ with $r_{C''}$ close to $r_{C'}$. At $r = r_{C''}$ a new period-doubling bifurcation occurs to a closed curve of period $8\,T$, and so on. The length of the successive intervals of r reduces in an apparently geometrical progression, so that the intervals entailing periods $2^n T$ with n large cannot be distinguished because of the limited graphical resolution of the plotter or the limited numerical resolution of the computer. At an accumulation point $r_\infty \approx 139.6$ the period becomes infinite, which means that the behavior actually is no longer periodic and it is called "aperiodic". This behavior is, however, not chaotic, because there is no sensitive dependence on the initial conditions.

If we further decrease the pump efficiency r to below $r_\infty \approx 139.6$, a continuously increasing quantity of chaotic behavior adds to the quasi-periodic one, so that the trajectory never closes on itself, and it fills an increasingly thicker region in the phase space. The attractor looks like a thick line or broad trajectory which seemingly closes on itself (Fig. 2.1.12). At the beginning, i.e. for r close to r_∞, the tube is thin and it closes only after many round trips. As r decreases further and the tube becomes thicker,

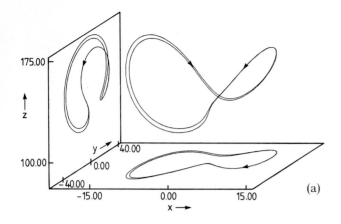

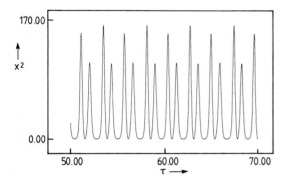

(a)

(b)

Fig. 2.1.10. The same as in Fig. 2.1.9, for $r = 143$. The period is $2T$.

the parts of the tube which are close to each other coalesce and the period associated to the broad trajectory decreases by a factor of 2. This "period dedoubling" or "period halving" occurs again and again, so that an "inverse" sequence or cascade of "noisy-period doublings" occurs: $\ldots \rightarrow 8T \rightarrow 4T \rightarrow 2T \rightarrow T$. Fig. 2.1.12 corresponds to the $2T$ noisy-period step of this cascade.

In fact, the inverse cascade's last step has such a large degree of chaotic behavior that in practice it is called or considered fully chaotic. This last step is reached at $r = r_B$, so that the attractor is no other than the chaotic one that we found for $r_A < r \leqslant r_B$ in Section 2.1.3.

Thus, in the present domain, $r_B < r < \infty$, we have found a sequence of "bifurcations" (i.e. qualitative changes in the characteristics of the attractor) which transforms a periodic attractor into a chaotic one. For this reason, it is known as a "road to chaos" and it is found in many non-linear dynamic systems. This and other known roads to chaos are described from a more general point of view in Chapter 3.

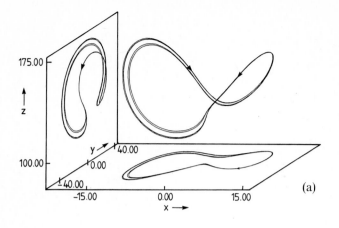

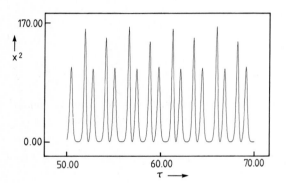

(a)

(b)

Fig. 2.1.11. The same as in Fig. 2.1.10, for $r = 140$. The period is $4\,T$.

2.2   The Logistic Map

2.2.1   Definition of the Map

Sometimes, for reasons inherent to the physical system or the detection equipment, it is not possible to obtain a continuous recording of a physical quantity. Instead, a sequence of measurements at discrete times, $x_1, x_2, ..., x_n, ...$, is obtained. This sequence defines a "map": $x_1 \rightarrow x_2 \rightarrow \cdots x_n$.

A map can also be theoretically defined by a function f which relates x_{n+1} with x_n for all n: $x_{n+1} = f(x_n)$. One of the simplest and better known non-linear maps is the "logistic" one, which is defined by

$$x_{n+1} = f(x_n) = \mu x_n (1 - x_n) \tag{2.2.1}$$

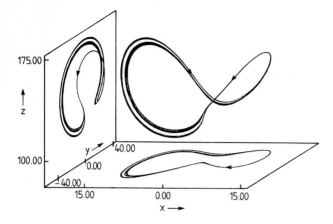

Fig. 2.1.12. The same as in Fig. 2.1.11.a, for $r = 139$. The "noisy" period is $2\,T$.

and is restricted to values $0 \leqslant x_n \leqslant 1$ for the variable and $0 < \mu < \infty$ for the parameter. This map has been used, for instance, to describe the evolution of a biological species in a bounded habitat (measured each year).

In spite of the fact that the logistic map does not correspond, in principle, to any kind of laser, it gives rise to a typical dynamic behavior which is also found in many non-linear systems and in particular in laser systems, including the Lorenz-Haken model analysed in the previous section. Let us consider this in some detail.

2.2.2   A Period-Doubling Sequence, again

Figure 2.2.1.a shows the values reached by the variable x_n for large n (i.e. the long-term values of the population, when the initial transient period has disappeared) as a function of the population growth rate μ. Several regimes can be distinguished, as follows.

For $0 \leqslant \mu \leqslant 1$ (not shown in the figure), the population growth rate is so low that no population exists: $x_n \to 0$. For $1 < \mu \leqslant \mu'$, the population evolves towards a stationary stable solution x_n shown by curve B. At $\mu = \mu'$, a "period-doubling" bifurcation appears: for $\mu' < \mu < \mu''$ the population evolves towards a periodic pulsing regime, where two values, given by the branches B_1' and B_2', alternate, so that the period is now 2 years. At $\mu = \mu''$, a new period-doubling bifurcation appears: for $\mu \geqslant \mu''$ the population reaches successively the values given by the branches B_1'', B_2'', B_3'' and B_4'', so that the sequence repeats every 4 years (period = 4 years). Another period doubling appears at a value $\mu = \mu'''$ slightly larger than μ'', and so on.

Thus, a road to chaos formed, as before, by a sequence of period doublings appears in the present case. At the accumulation point $\mu^{(\infty)} = 3.57$, the period is infinite, so that the evolution is aperiodic. Also as before, for $\mu > \mu^{(\infty)}$ an inverse sequence of noisy-period doublings appears, which ends at $\mu = \bar{\mu}'$ where a fully chaotic evolution

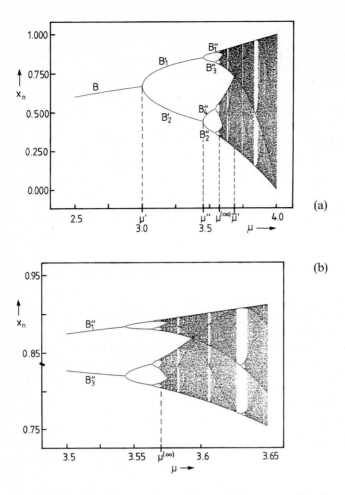

Fig. 2.2.1. The logistic map. Dependence of the asymptotic values of x_n (i.e. for large n) on the parameter μ. (a) General view of the unstable domain; (b) a blow-up of the region close to B_1'' and B_3''.

appears. A sensitive dependence on the initial conditions is again one of the most significant features of this chaos. Periodic windows of periods 3 years, 5 years and others are apparent within the chaos (see the vertical white strips). Notice the self-similarity property of the figure: if some small zones are magnified (e.g. Fig. 2.2.1.b), again a period-doubling sequence appears. Period-doubling sequences appear even on one side of each periodic window.

The period-doubling sequences seen in the Lorenz model (Section 2.1.4) and in the logistic map are identical in many features. This similarity would appear even more evident if in the Lorenz case we had plotted the Poincaré maps (defined in Fig. 2.1.5) for trajectories in phase space corresponding to each step of the period-doubling sequence (examples of these trajectories are shown in Figs. 2.1.9 − 2.1.11): they would be constituted by a unique pair of points for $r_C \leqslant r < \infty$ (intersection of the trajectory of Fig. 2.1.9 with the plane $z = r - 1$), two pairs of points for $r_{C'} \leqslant r < r_C$ (Fig. 2.1.10), four pairs of points for $r_{C''} \leqslant r < r_{C'}$ (Fig. 2.1.11), and so on. By plotting the distance between each pair of points as a function of r, a figure very similar to Fig. 2.2.1 would result.

This analogy in the behavior of two very different systems for limited ranges of values of some parameter reveals that the time evolution of non-linear systems, which has always been considered to be extraordinarily complex, can present common features, several of which are at present fairly well understood and are described in simple terms in Chapter 3. Before going into this subject, we give below a graphical explanation of the origin of period-doubling bifurcations appearing in the logistic map. The reader who is not interested in this aspect may go directly to Chapter 3.

2.2.3 Origin of the Period-Doubling Sequence

The logistic map (2.2.1) can be graphically represented by plotting x_{n+1} versus x_n, as in Fig. 2.2.2, Fig. 2.2.2.a corresponds to a value of μ smaller than μ', for which the map tends asymptotically towards a unique value (see Fig. 2.2.1.a) whatever the initial value x_1 may be. Figure 2.2.2.b corresponds to a value of μ satisfying $\mu' \leqslant \mu < \mu''$, for which any sequence eventually oscillates between two values given by the branches B'_1 and B'_2 in Fig. 2.2.1. Comparison of Figs. 2.2.2.a and b shows that the period-

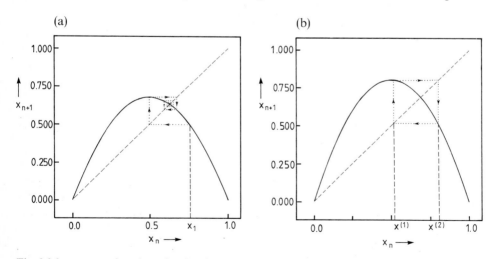

Fig. 2.2.2. x_{n+1} as a function of x_n for the logistic map. The dotted line represents an example of a sequence $x_1 \rightarrow x_2 \rightarrow x_3 \rightarrow \dots$ (a) Case with $\mu = 2.7$, in which the sequence initiating at x_1 tends towards a unique value $x = 1 - 1/\mu$ for large n. (b) Case with $\mu = 3.2$, in which the sequence shown oscillates permanently between two values $x^{(1)}$ and $x^{(2)}$.

doubling bifurcation at $\mu = \mu'$ arises when the modulus of the slope of the curve at the intersection point with the diagonal line (dashed line) becomes larger than unity.

The next period-doubling bifurcation appears in a similar way: at the bifurcation point μ'', the map $f^{(2)}(x_n)$ (where $f^{(n)}$ denotes the application n times of the map f) undergoes a change in its slope at the intersection point with the diagonal identical with that experienced by the map f at $\mu = \mu'$. The same occurs at the next bifurcation point μ''' with the map $f^{(4)}(x_n)$, and so on. In this way a complete period-doubling sequence appears. The simplicity of this interpretation explains why the period-doubling road to chaos is so often encountered in non-linear dynamic systems.

3 Non-linear Dynamic Behavior: Mathematical Concepts

After the illustrative examples shown in the preceding chapter, we proceed to decribe here, from a mathematical point of view and in simple terms, the general features that can be expected in non-linear dynamic behavior. This will be applied in subsequent chapters to the study of the dynamics of the different kinds of lasers.

3.1 Non-Linear Differential Equations and Maps

3.1.1 Non-Linear Differential Equations

Any kind of laser can be described, with some simplifications, by means of a set of coupled non-linear differential equations involving first-order time derivatives. Spatial dependence, of course, also exists, but so far it has been analysed with less detail and will not be considered here (it is only taken into account in some specific sections in subsequent chapters). Thus, in general we have to deal with equations such as

$$\frac{d\vec{x}}{dt} = F_{\vec{\mu}}(\vec{x}, t) \tag{3.1.1}$$

where the time-dependent vector $\vec{x} = (x_1, ..., x_n)$ represents the n dynamical variables describing the laser system, so that its evolution defines a "trajectory" or "orbit" of the system in the "phase space" defined by these variables. The smooth (i.e. differentiable) function or vector field $F_{\vec{\mu}}$ describes the non-linear coupling existing between the dynamical variables in a given kind of laser. In general, $F_{\vec{\mu}}$ depends on several "control parameters" $\mu_1, ..., \mu_p$ that characterize each specific set of experimental conditions; this set of parameters is designated by a vector $\vec{\mu} = (\mu_1, ..., \mu_p)$.

For instance, for the Lorenz model described in Section 2.1.1 (see eqns. (2.1.1)), the dynamical variables are x, y and z, the control parameters are σ, b and r, and the function $F_{\hat{\mu}}$ is given by the right hand terms of eqns. (2.1.1). Figures 2.1.1.a, 2.1.2.a and 2.1.3.a are examples of trajectories or orbits in the phase space.

When the function F does not explicitly depend on time, the system is called "autonomous":

$$\frac{d\vec{x}}{dt} = F(\vec{x}) \tag{3.1.2}$$

where the subscript $\hat{\mu}$ on F has been omitted for simplicity. Lasers are autonomous systems, except when a time-modulated external influence is introduced. Unless it is explicitly indicated, in this chapter autonomous systems are considered.

Given an initial condition $\vec{x}(t = 0) = \vec{x}^0$, it is well known that equations such as eq. (3.1.2) have a unique solution. This means that trajectories in the phase space never intersect. The only exceptions to this rule are the "singular" points, at which several trajectories can eventually end (or begin); the different kinds of singular points (or orbits) are described in Section 3.2.1 (see also table 3.3.1 and 3.4.1).

A solution $\vec{x}(t)$ of eq. (3.1.2) with initial condition $\vec{x}(0) = \vec{x}^0$ is sometimes represented by means of a time evolution operator $\mathscr{F}^t$, which when applied to $\vec{x}^0$ gives the point $\vec{x}(t)$ reached by the system at time t:

$$\mathscr{F}^t(\vec{x}^0) = \vec{x}(t)\,. \tag{3.1.3}$$

Lasers are "dissipative systems", i.e. they have energy exchanges with a thermal bath or a large system with many degrees of freedom. Dissipation or internal friction has an important consequence for autonomous systems: any volume V in the phase space contracts along time evolution:

$$\mathscr{F}^t(V) < V, \quad \text{for all} \quad t > 0\,. \tag{3.1.4}$$

See as an example eq. (2.1.4) for the Lorenz model. All the dynamic behaviors considered in this book refer to dissipative systems.

3.1.2 Reduction Methods

From the theoretical point of view, the study of a specific laser system requires the analytical or, in general, numerical solution of a differential equation such as eq. (3.1.2) for many different initial conditions and experimental conditions. In many cases this can be an excessively large quantity of information to handle or understand. Therefore, methods or techniques for reducing the amount of information or the number of degrees of freedom, in such a way that the basic features of the dynamical evolution be retained, are very useful in these cases.

An almost equivalent problem appears from the experimental point of view: it is usually impossible to measure or to deal with the complete temporal evolution of all the dynamic variables, and only partial information is recorded, which should also keep the basic dynamic features.

Here we briefly describe or comment on some of the most often used reduction methods.

a. Low-dimensional projections

Often in experiments only the temporal evolution of some of the dynamic variables can be detected. This corresponds to observing only a projection of the orbit in the phase space on the subspace defined by these variables. On the other hand, one, two- or three-dimensional projections of the calculated orbits are also used in theoretical work, as for instance in Fig. 2.1.4.b for the Lorenz model.

If only a few variables are recorded, obviously part of the physical information about the system is lost. However, because the non-linear terms in eqns. (3.1.2) couple together all the variables, the dynamic features characterizing the time evolution can be observed on any one of the variables. For instance, it is sufficient to look at only one variable to obtain evidence for a stationary, periodic or chaotic time evolution. Based on this fact, in the next chapter a way of characterizing a dynamic behavior from a time series of measurements on any variable of the system is presented.

b. Maps, Poincaré map

A different way of obtaining partial but substantial information about the dynamic behavior consists in calculating or measuring the value of the variables only at a discrete sequence of times t_m ($m = 0,1,2,...$). In this way a "map" f is defined:

$$\vec{x}_{m+1} = f(\vec{x}_m) \tag{3.1.5}$$

where $\vec{x}_m = \vec{x}(t_m)$.

A useful kind of map is the "Poincaré map", which is defined by the sequence of intersection points of an orbit γ in the phase space (say, for instance, a three-dimensional phase space) with a given surface Σ ("surface of section"), as illustrated in Fig. 3.1.1. Only the intersection points corresponding to a given crossing direction (for instance, from left to right in the figure) must be considered. Note that the time intervals separating two consecutive points in a Poincaré map are not necessarily constant.

Poincaré maps are often used with systems ruled by three equations, where they represent a reduction of the flow to a two-dimensional map. Figure 2.1.5.b for the Lorenz model is an example of such a kind of Poincaré map.

In addition to the drastic reduction in the amount of information to be dealt with, the interest in the Poincaré maps lies in the fact that they display the same kind of topological properties as the complete solutions of eq. (3.1.2), so that their analysis reveals most of the properties of these solutions.

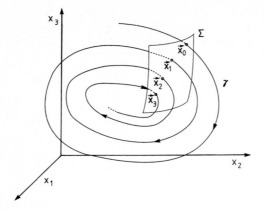

Fig. 3.1.1. Poincaré map $\vec{x}_0 \rightarrow \vec{x}_1 \rightarrow \vec{x}_2 \rightarrow \vec{x}_3 \rightarrow \dots$ defined by the intersection points between an orbit γ and a surface of section Σ in the phase space.

Another kind of map is obtained when a variable is recorded or calculated at discrete times $t_m = m(\Delta t)$ ("stroboscopic" study). In certain cases a description through a map is the only one possible, because a set of non-linear differential equations such as (3.1.2) cannot be defined or is not known; an example is the logistic map described in Section 2.2.

c. Adiabatic elimination of variables (Slaving principle)

Often, in equations such as (3.1.2), one or several of the control parameters μ_j influence in a very direct way the temporal evolution of particular dynamical variables. For instance, in the Lorenz laser equations (3.1.1), the normalized field and population relaxation rates σ and b influence, respectively, the field and population variables x and z, through the terms $-\sigma x$ and $-bz$ appearing in the respective evolution equations $dx/d\tau$ and $dz/d\tau$. In these cases, i.e. when a control parameter μ_j influences the time evolution of a variable x_j through a term $-\mu_j x_j$, the following behavior can occur. If this parameter is much larger than the remaining ones, the influenced variable x_j rapidly "loses" the memory of its history (i.e. of the values reached at preceding times), in such a way that it rapidly adapts to (or "adiabatically follows") the instantaneous values reached by the remaining variables, with an "adaptation speed" approximately proportional to μ_j. Hence a good simplification for solving eq. (3.1.2) within a given time scale (or time resolution) consists in equating to zero, for all t, the time derivatives corresponding to the variables relaxing within a faster time scale than the given one, and introducing the resulting algebraic equations into the remaining differential equations. In this way, the number of independent equations and of independent variables or degrees of freedom are reduced ("adiabatic elimination").

Since, under these conditions, the slowly evolving variables completely determine the evolution of the physical system, they are called the "order parameters". It is said that they "slave" the subsystems controlled by the fast variables. Indeed, when a few order parameters rule the time evolution of a large system with many degrees of freedom, it is a sure sign that a large degree of order has been reached. Lasers are typical systems where the "slaving principle" applies, as Haken has elegantly explained (Haken, 1983 and 1985, 3rd ed.).

3.2 Dynamical Stability

It is well known that in a physical experiment it is impossible to reproduce exactly any given initial conditions, or to avoid completely the presence of some noise or fluctuations which "contaminate" the signals or quantities to be measured. Hence it is very important to study the influence of small variations in the initial conditions on the solutions of eq. (3.1.2). The sensitivity of a given solution to these variations defines its "dynamical stability".

We deal in this section with the dynamical stability of two important classes of solutions of differential equations such as (3.1.2): stationary and periodic solutions (Iooss and Joseph, 1980; Guckenheimer and Holmes, 1983).

3.2.1 Stability of Stationary and Periodic Solutions. Basic Concepts

a. Stationary Solutions

The stationary solutions of eq. (3.1.2) are those satisfying

$$\frac{d\vec{x}}{dt} = F(\vec{x}) = 0 \tag{3.2.1}$$

so that they remain constant with time:

$$\vec{x}(t) = \vec{x}^0 . \tag{3.2.2}$$

For this reason they are also known as "fixed points" (in the phase space).

With respect to the stability of a fixed point, three kinds of behaviors can be distinguished:

(a) A stationary solution or fixed point $\vec{x}^0$ is "stable" (or a "sink", or "asymptotically stable") if any other solution originating close to $\vec{x}^0$ approaches it asymptotically or, in other words, when any small perturbation added to the stationary solution damps with time. Depending on the way in which the perturbation damps, two kinds of stable fixed points are distinguished: "node" (Fig. 3.2.1) and "focus" (Fig. 3.2.2). The points

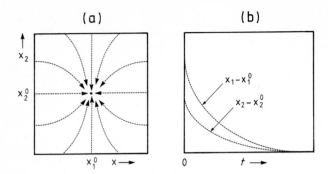

Fig. 3.2.1. (a) A node (x_1^0, x_2^0) and several trajectories (dashed line) asymptotic to it. (b) Evolution with time of the variables x_1 and x_2 corresponding to one of these trajectories.

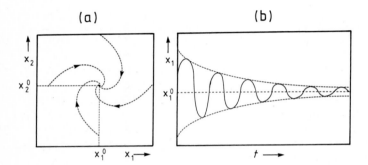

Fig. 3.2.2. (a) A focus (x_1^0, x_2^0) and several trajectories (dashed line) asymptotic to it. (b) Evolution with time of one of the variables corresponding to one of these trajectories.

C_+ and C_- in Figs. 2.1.1 − 2.1.3 (corresponding to the Lorenz model) are examples of a focus.

(b) A fixed point $\vec{x}^0$ is "marginally stable" (or a "center", or "neutrally stable") if any other solution originating close to $\vec{x}^0$ remains there forever, i.e. when any small perturbation neither damps nor amplifies with time.

(c) Finally, a fixed point $\vec{x}^0$ is "unstable" if it is neither stable nor marginally stable. If any trajectory originating close to $\vec{x}^0$ goes away (as, for instance, in Figs. 3.2.1 and 3.2.2 if the arrows are drawn in the opposite direction), it is called a "source", but if some of the trajectories (for instance, those going along a certain direction) converge towards $\vec{x}^0$, then it is known as a "saddle point". Figure 3.2.3 shows an example of a saddle-point in a two-dimensional phase space.

In relation to the stability of a fixed point, some further concepts are useful. The "local stable manifold", W_{loc}^s, of a fixed point $\vec{x}^0$ is a set of points close to $\vec{x}^0$ which

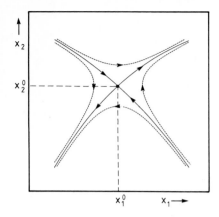

Fig. 3.2.3. A saddle point or saddle node (x_1^0, x_2^0). Continuous line: converging and diverging trajectories. Any other trajectory (dashed line) is divergent.

satisfy (i) the trajectories originating at any of these points asymptotically approach $\bar{x}^0$ when $t \rightarrow \infty$; (ii) these trajectories remain for all time ($t \geqslant 0$) within W_{loc}^s.

The "global stable manifold", W^s, of a fixed point $\bar{x}^0$ is obtained by backward dynamical evolution of W_{loc}^s (or, equivalently, by eliminating the condition of proximity to $\bar{x}^0$ in the given definition of a local stable manifold).

In an equivalent way, the "local unstable manifold", W_{loc}^u, and the "global unstable manifold", W^u, of a fixed point $\bar{x}^0$ can be defined. This simply requires changing the sign of all the time evolutions mentioned in the above definitions (for instance, in the condition (i) above, $t \rightarrow -\infty$ must replace $t \rightarrow \infty$).

A sink (source) point has no local unstable (stable) manifolds, but a saddle point has both stable and unstable manifolds. For instance, for the saddle point of Fig. 3.2.3, W_{loc}^s and W_{loc}^u are given by the converging and diverging trajectories (continuous line), respectively.

The trajectories shown in Figs. 2.1.1 – 2.1.3 (Lorenz-Haken model) represent the global unstable manifold of the zero point (they also belong to the global stable manifold of the points C_+ and C_-). This manifold is one-dimensional (i.e. it is a curve), but the global stable manifold of the same point is two-dimensional (i.e. a surface), as Fig. 3.2.4 shows. The zero point in the Lorenz model is, therefore, a saddle point (or, more specifically, a saddle-node) in a three-dimensional phase space.

The uniqueness property of the solutions of eq. (3.1.2) implies that two stable (or unstable) manifolds of distinct fixed points cannot intersect, nor can W^s (or W^u) intersect itself. However, intersections or coincidences can occur between stable and unstable manifolds of the same or distinct fixed points; examples of this are the "homoclinic" and "heteroclinic" orbits, which are defined in Section 3.3.3.b. In many cases the global stable and unstable manifolds of the same or different fixed points present very intricate shapes; they can fold and twist in very complex manners, approaching themselves arbitrarily closely.

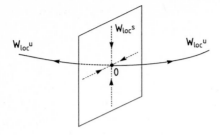

Fig. 3.2.4. Local stable and unstable manifolds
of the zero point, in the Lorenz model, for $r > 1$ (see Section 2.1).

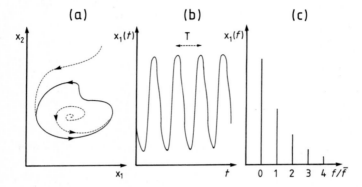

Fig. 3.2.5. (a) A periodic orbit (continuous line) and some trajectories (dashed line) asymptotic to it. (b) Evolution with time of one of the variables, for the periodic orbit; T is the period. (c) Fourier power spectrum associated with the same variable; $\bar{f} = 1/T$ is the fundamental frequency. The presence of the harmonics $2\bar{f}, 3\bar{f}, \ldots$ indicates that the temporal oscillation is not sinusoidal.

b. Periodic Solutions

A "periodic orbit", "closed orbit" or "limit cycle" is a solution of eq. (3.1.2) that satisfies

$$\vec{x}(t + T) = \vec{x}(t) \tag{3.2.3}$$

for all t and $0 < T < \infty$, where T represents the period. Figure 3.2.5 shows an example of a simple periodic orbit. The dynamic periodicity is clearly seen in both the time dependence (b) and frequency spectrum of any of the variables (c). Figures 2.1.8 – 2.1.11 in the preceding chapter represent more complex periodic orbits.

The concepts of stability, marginal stability and instability are identical with those given above for a fixed point, and the local and global stable or unstable manifolds are also defined in the same way. The periodic orbit shown in Fig. 3.2.5.a is stable, and by changing the directions of all the arrows it transforms into an unstable orbit.

3.2.2 Methods for Stability Analysis. Characteristic Exponents

a. Linear Stability Analysis

Because the stability behavior of a stationary or periodic solution is related to the time evolution of small perturbations, it can be easily determined through a linearization of the set of equations (3.1.2) around the stationary or periodic solution. This method is known as the "linear stability analysis", which in the case of a periodic solution gives rise to the "Floquet theory" (Iooss and Joseph, 1980).

Let $\delta \vec{x}(0)$ be a small perturbation added at $t = 0$ to a stationary or periodic solution $\vec{x}^0(t)$ ($\vec{x}^0(t) = \vec{x}^0$ in the stationary case and $\vec{x}^0(t) = \vec{x}^0(t + T)$ in the periodic case); see Fig. 3.2.6. The time evolution of the perturbation, $\delta \vec{x}(t)$, for $t \geqslant 0$ can be found by introducing the whole solution

$$\vec{x}(t) = \vec{x}^0(t) + \delta \vec{x}(t) \tag{3.2.4}$$

into eq. (3.1.2). If the resulting set of equations is linearized with respect to $\delta \vec{x}(t)$ (i.e. all the non-linear terms are ignored), it yields

$$\frac{\mathrm{d}}{\mathrm{d}t} \delta \vec{x}(t) = L(t) \delta \vec{x}(t) \tag{3.2.5}$$

where $L(t) = [\partial F_i / \partial x_j]_{\vec{x} = \vec{x}^0(t)}$ represents the $n \times n$ matrix of the partial derivatives of the function F of eq. (3.1.2) with respect to the variables x_n ($\mathrm{d}x_i/\mathrm{d}t = F_i(x_1,\ldots,x_j,\ldots,x_n)$), performed at the point $\vec{x} = \vec{x}^0(t)$. The matrix $L(t)$ is constant in the case of a stationary solution, whereas it is periodic (with period T) in the case of a periodic orbit. The set of linear differential equations (3.2.5) has a linearly independent set of n complex solutions of the form

$$\delta \vec{x}^{(i)}(t) = e^{\lambda_i t} \vec{v}^{(i)}(t) \tag{3.2.6}$$

where $i = 1,\ldots,n$ denotes each one of the solutions; λ_i represents the corresponding "characteristic exponent", which has a multiplicity m_i that satisfies $0 < m_i \leqslant n$, $\sum_{i=1}^{n} m_i = n$, and the vectors $\vec{v}^{(i)}(t)$ are polynomials in t whose highest power is smaller than m_i:

$$\vec{v}^{(i)}(t) = \vec{v}_0^{(i)}(t) + \vec{v}_1^{(i)}(t)t + \cdots + \vec{v}_{m_i-1}^{(i)}(t)t^{m_i-1} . \tag{3.2.7}$$

In the fixed-point case, the characteristic exponents λ_i and the vectors $\vec{v}^{(i)}$ are respectively the eigenvalues and associated eigenvectors of the constant matrix L, whereas in the closed-orbit case the λ_i are the eigenvalues of a matrix Λ which defines the time periodicity of any solution of eq. (3.2.5):

$$\delta\vec{x}^{(i)}(t + T) = e^{AT}\delta\vec{x}^{(i)}(t) \tag{3.2.8}$$

and $\vec{v}^{(i)}(t)$ are T-periodic vectors. The matrix e^{AT} is known as the Floquet matrix.

From the real and imaginary parts of the complex solutions (3.2.6), a set of n independent real (and therefore physically meaningful) solutions can be extracted, which form a basis for any other real solution of eq. (3.2.5), in particular for the solution satisfying the required initial conditions.

Hence the temporal evolution of the initial perturbation $\delta\vec{x}(0)$ is basically ruled by the characteristic exponents λ_i. In particular, $\text{Re}\{\lambda_i\}$ determines the stability behavior of the fixed-point or periodic orbit: if the real parts of all the λ_i are negative then $\vec{x}^0(t)$ is stable, whereas if at least one of these real parts is positive then it is unstable (in the case of a closed orbit there is always an exponent equal to zero, which corresponds to a perturbation along the closed orbit).

If one is interested, additional information can be obtained from the characteristic exponents λ_i and vectors $\vec{v}^{(i)}(t)$. For instance, in the case of a fixed point $|\text{Re}\{\lambda_i\}|$ represents the rate at which a small initial perturbation $\vec{v}^{(i)}(0) + \vec{v}^{(i)}(0)^*$ increases $(\text{Re}\{\lambda_i\} > 0)$ or decreases $(\text{Re}\{\lambda_i\} < 0)$ with time; $|\text{Im}\{\lambda_i\}|$ is the angular frequency at which a small initial perturbation $\vec{v}^{(i)}(0) + \vec{v}^{(i)}(0)^*$ spirals around a stable or unstable focus $\vec{x}^0$ (see Fig. 3.2.2); if $\text{Im}\{\lambda_i\} = 0$, then the fixed point is a node (Fig. 3.2.1) instead of a focus; finally, $\vec{v}^{(i)}(0) + \vec{v}^{(i)}(0)^*$ represents the perturbation whose temporal evolution is exclusively ruled by the exponent λ_i.

The stability behavior of a closed orbit can also be characterized through the eigenvalues ξ_i of the Floquet matrix $\exp\{AT\}$ of eq. (3.2.8), which are known as the "Floquet multipliers". The closed orbit is stable if all the multipliers lie inside a circle of radius unity in the complex plane (except one of them which is strictly equal to unity), whereas it is unstable if at least one multiplier lies outside the circle.

b. Weak Side-Band Approach

The stability behavior of a fixed point can also be analysed through another linear approach, which is mathematically similar to the previous one but the physical interpretation of the main steps differs.

Because the perturbation $\delta\vec{x}(t)$ in eq. (3.2.4) spirals or rotates around the fixed point $\vec{x}^0$ (see Fig. 3.2.6.a), it can be regarded as a pair of small-amplitude complex side-bands that modulate the constant "signal" $\vec{x}^0$:

$$\vec{x}(t) = \vec{x}^0 + \delta\vec{x}(0)e^{-i\omega t} + \delta\vec{x}(0)^*e^{i\omega t}. \tag{3.2.9}$$

This expression is in fact the same as eq. (3.2.4) if one selects there a perburbation of the kind given by eq. (3.2.6), i.e. a perturbation $\delta\vec{x}(0) + \delta\vec{x}(0)^*$ whose time evolution is given by $\exp\{\lambda_i t\}$, with $\lambda_i = -i\omega$.

The stability or instability criteria can be now imposed through the gain and phase-matching-conditions. Instability requires, on one hand, the side-band gain to exceed losses, and on the other, the side-bands to phase match the "boundary" conditions of

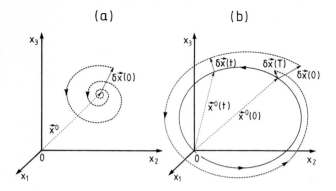

Fig. 3.2.6. Examples of evolution with time of a perturbation $\delta \vec{x}(0)$ (dashed line) added to (a) a stationary solution or fixed point $\vec{x}^0$ or (b) a periodic orbit $\vec{x}^0(t)$. $\delta \vec{x}(T)$ is the value of the perturbation after a period T. In both examples the perturbation damps with time.

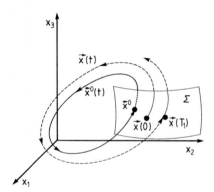

Fig. 3.2.7. A closed orbit $\vec{x}^0(t)$ of period T and a nearing trajectory $\vec{x}(t)$. The intersection of this trajectory with the surface of section Σ defines a Poincaré map $\vec{x}(0) \rightarrow \vec{x}(T_1) \rightarrow \ldots$

the system. If, when imposing these conditions, all the non-linear terms with respect to the amplitude $\delta \vec{x}(0)$ are neglected, the results obtained are in general equivalent to those provided by the linear stability analysis. A detailed comparison is given by Abraham et al. (1985). The standard linear stability analysis seems more adequate for further mathematical analysis than the weak side-band approach.

c. Poincaré Map Method

This method is especially suited to studying the stability behavior of a closed orbit. Let us choose a surface of section Σ (see Section 3.1.2.b) that intersects the closed orbit $\vec{x}^0(t)$ (or period T) at a point $\vec{x}^0$ (Fig. 3.2.7). The stability behavior can be determined by looking only at the intersection points with Σ of any nearing trajectory $\vec{x}(t)$, which define a Poincaré map:

$$\vec{x}(0) \rightarrow \vec{x}(T_1) \rightarrow \cdots \rightarrow \vec{x}(T_m) \rightarrow \cdots \tag{3.2.10}$$

where T_m ($m = 1,2,\dots$) are the time intervals separating two consecutive intersections of the trajectory with Σ. In general, these intervals are close to T, and satisfy $T_m \rightarrow T$ when $\vec{x}(T_m) \rightarrow \vec{x}^0$. By denoting $\vec{x}(T_m) = \vec{x}_m$, the map can be expressed as eq. (3.1.5).

Stability of the closed orbit requires the sequence of moduli of the points in any Poincaré map such as (3.2.10) originating close to $\vec{x}^0$ to go to zero, whereas instability requires that at least for one of the maps the sequence be divergent. In addition to the numerical direct method of verifying these criteria, there is also a linear method which uses the Floquet theory to study the properties of the map: if the basis of solutions (3.2.6) is conveniently chosen, then the last column of the Floquet matrix in (3.2.8) is $(0,\dots0,1)^T$, and the matrix defining the linearized Poincaré map is simply the $(n - 1) \times (n - 1)$ matrix obtained by deleting the nth row and column of the Floquet matrix. Thus, the eigenvalues of the Poincaré map coincide with the first $n - 1$ Floquet multipliers and the stability behavior can be determined as described above.

d. Stability of a General Solution. Lyapunov Exponents

When a solution is not stationary or periodic, the study of its sensitivity to small variations in the initial conditions cannot be performed, in general, by means of simple analytical theories. Usually one has to calculate numerically or measure the time evolution of a given solution and of any other close trajectory. Characteristic exponents such as those defined through eqns. (3.2.6) or (3.2.8) do not exist, but other (real) ones can be established which have a meaning similar to that of their real parts. They are known as the "Lyapunov exponents" and are defined as

$$\lambda_i = \underset{t \rightarrow \infty}{\text{Lim sup}} \left\{ \frac{1}{t} \ln | \delta\vec{x}^{(i)}(t) | \right\} \tag{3.2.11}$$

where, similarly to (3.2.4), $\delta\vec{x}^{(i)}(t)$ represents the time evolution of a small perturbation added to a given solution $\vec{x}(t)$; the superscript i ($i = 1,\dots,n$) denotes each one of a set of n linearly independent perturbations that can be choosen. Lim sup($t \rightarrow \infty$) of an infinite set of real numbers $\{\dots\}$ denotes the infimum of all numbers r with the property that only a finite subset of $\{\dots\}$ exceeds r; this kind of limit allows one to distinguish between a short-term and a long-term evolution, which actually defines the stability of a trajectory.

A solution is stable whenever all the λ_i are negative (or zero) and it is unstable when at least one of them is positive. In the case of a stationary or a periodic solution, the Lyapunov exponents coincide with the real part of the characteristic exponents λ_i defined through eq. (3.2.6). For instance, for a system described by three variables (three-dimensional phase space), the Lyapunov exponents $(\lambda_1, \lambda_2, \lambda_3)$ corresponding to stable solutions take the following signs:

Fixed point: $(-,-,-)$
Closed orbit: $(-,-,0)$

3.3   Asymptotic Behavior: Attractors

3.3.1   Attractors: Definition and Properties

For dissipative systems any solution or trajectory can be divided into two parts: the initial transient regime and the final or asymptotic regime, which is located within an "attractor".

An "attractor" is a set of points in the phase space towards which trajectories eventually tend. We have found several examples of attractors in preceding sections: the points C_+ or C_- in Figs. 2.1.1 to 2.1.3, the "strange" attractors of Figs. 2.1.4 and 2.1.12, the closed curves of Figs. 2.1.8 to 2.1.11, the stable fixed points of Figs. 3.2.1, 3.2.2 and 3.2.6.a and the closed curves of Figs. 3.2.5.a and 3.2.6.b. In the case of fixed points and closed orbits, it is obvious that only the stable ones can define attractors.

Because an attractor is a basic concept for characterizing the asymptotic dynamic behavior in dissipative systems, precise mathematical definitions have been formulated (Guckenheimer and Holmes, 1983; Eckmann and Ruelle, 1985). Explained in simple terms, these definitions include four basic points or conditions for a (compact) set A to be an attractor:

(i) The neighborhood of A contracts, with time, towards A ("shrinking neighborhood").

(ii) Any trajectory originating inside the attractor remains within it ($\mathscr{F}^t(A) = A$ for all t).

(iii) Any trajectory within the attractor goes through all of it: for arbitrarily large values of t the trajectory passes arbitrarily close to any point of the attractor (recurrence property).

(iv) It is non-decomposable: it cannot be split into two disjoint pieces each satisfying the preceding conditions.

Many different trajectories can eventually settle on a given attractor. The set of points in the phase space which evolve towards a given attractor constitute its "basin of attraction".

There is no limit to the number of attractors that can exist for a given dynamical system (usually this number is small, but some simple systems have an infinity of them). The basins of attraction may be complicated, even if the attractors are simple; they may interlace in almost inextricable manners, defining "fractal" boundaries (see Section 2.1.3 and Fig. 2.1.6 for the fractal concept).

"Repellors" can also be defined; these have the opposite meaning to that of the attractors. They are sets of points from which trajectories go away. Physically they are less relevant than attractors.

The contraction of volumes in phase space (see eq. (3.1.4)) confers a common and important feature to any attractor: its volume is zero. This means that the dimension of an attractor is lower than that of the phase space. This property is useful for classifying the different kinds of attractors, which is done in the next section.

3.3.2 Kinds of Attractors

We give here a simple classification of attractors based only on their dimension and other most apparent features.

a. One- and Two-Dimensional Phase Spaces

For systems described by only one or two variables (let us call them 1-D or 2-D cases), the general properties described in Section 3.1.1 and the Poincaré-Bendixon theorem (Hirsch and Smale, 1974) reduce the possible kinds of attractors to only a few simple ones (see the summary in Table 3.3.1):

For 1-D cases, they coincide with the stable fixed points, defined in Section 3.2.1.a. For 2-D cases, they coincide with the stable fixed points and stable closed orbits (Section 3.2.1.b). Figures 3.2.1.a, 3.2.2.a and 3.2.5.a show typical examples of attractors.

b. Three- and Higher-Dimensional Phase Spaces: Periodic and Quasi-Periodic Attractors

The periodic attractors appearing in 3-D and n-D ($n > 3$) phase spaces are generalizations of the previous ones. Fixed points and closed orbits obviously exist, but the last ones are not necessarily confined on a plane.

Table 3.3.1. Simple schematic classification of attractors, for the different possible dimensions (n-D) of the phase space ($n = 1,2,3,...$); see the main text for additional information.

	Fixed points	Periodic orbits	Quasi-periodic attractors	Strange fractal	chaotic
1-D	Yes	No	No	No	No
2-D	Node, focus	Closed curves	No	No	No
3-D	Node, focus, ...	Closed curves	Closed surfaces	Yes	Yes
n-D	Node, focus, ...	Closed curves	Closed surfaces Closed manifolds (dim $m < n$)	Yes	Yes

The next higher-order attractors are closed surfaces (or "two-dimensional mani-folds") (see Table 3.3.1). The simplest is a toroidal surface or torus, T^2, in a 3-D phase space (Fig. 3.3.1); it appears when the time evolution is doubly periodic, for instance, when the trajectories can be described through two variables, such as y_1 and y_2 in Fig. 3.3.1, whose time dependence is periodic, with respective periods T_1 and T_2:

$$y_i(t) = \frac{2\pi}{T_i} t \quad (\text{mod } 2\pi). \tag{3.3.1}$$

Two cases may be distinguished. First, if T_2/T_1 is a rational number, then the trajectory is a closed line on the toroidal surface (see the example in Fig. 3.3.1, for which T_2/T_1 = 1/4). In fact, in this case the attractor is not the toroidal surface but only the closed line. If the time evolution along this orbit is described through the phase space variables $x_i(t)$ (i = 1,2 or 3), their Fourier spectra contain both frequencies $f_1 = 1/T_1$ and $f_2 = 1/T_2$ (and also, in general, any linear combination $lf_1 + mf_2$, where l and m are integers (positive or negative), if the periodic evolution is not exactly sinusoidal) (Fig. 3.3.2).

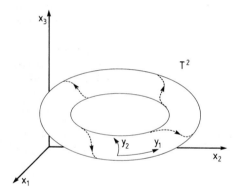

Fig. 3.3.1. A two-dimensional attractor in a 3-D phase space: a toroidal surface or torus T^2. y_1 and y_2 define a set of coordinates on the torus and the dashed line represents a periodic orbit on it.

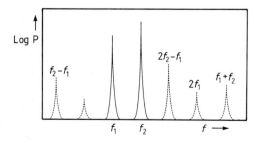

Fig. 3.3.2. Schematic representation of the power spectrum $P(f)$ of a closed orbit with a double periodicity at frequencies f_1 and f_2, as for instance a closed orbit on the toroidal surface in Fig. 3.3.1.

The other possible case is when T_2/T_1 is an irrational number (i.e. the periods T_1 and T_2 are incommensurate). The trajectory covers the whole toroidal surface and does not close (i.e. it does not reach the initial point) for any finite time t. In this case the trajectory is periodic in each of its coordinates y_1 and y_2, but there is no periodicity for the whole evolution (or it is "periodic" with a period $T \to \infty$); we are in the presence of a "quasi-periodic attractor", which is constituted by the toroidal surface (or, in general, by a two-dimensional manifold).

It might seem an unphysical refinement to distinguish between rational and irrational numbers for the ratio T_2/T_1, if one takes into account experimental uncertainty. See for this aspect Bergé, Pomeau and Vidal (1984), where the locking effect between two oscillators is considered.

The conclusions of this illustrative example can be easily generalized to phase spaces of any dimension $n \geqslant 3$: the periodic orbits can display from one- to $(n - 1)$-fold periodicity, and the tori and "hyper-tori" $T^{n'}$ can have dimensions n' from 2 to $(n - 1)$. The hyper-torus dimension indicates the number of different periods ruling the corresponding quasi-periodic dynamic behavior. In general, periodic and quasi-periodic attractors in an n-dimensional phase space are m-dimensional closed manifolds, with $m < n$ (see Table 3.3.1).

c. "Strange" Attractors

In n-dimensional phase spaces, with $n \geqslant 3$, attractors can exist that are not manifolds (i.e. smooth curves, surfaces, hyper-surfaces, etc.) and do not describe periodic or quasi-periodic dynamic evolution. They are known as "strange" attractors, which means that they are "fractal" and "chaotic". We have already found a typical example of such a kind of attractor in the Lorenz model: Fig. 2.1.4 shows a trajectory captured by a strange attractor. Since in Section 2.1.3, where Fig. 2.1.4 is analysed, we have already given detailed and intuitive explanations about the meaning of the terms "fractal" and "chaotic", here we only briefly recall it.

A fractal attractor is characterized by the fact that its Hausdorff dimension (which is defined in Section 4.2.2) is larger than its geometrical (or topological) one and usually is not an integer. Its origin lies in the intense stretching and repeated folding in some directions during the volume contraction in phase space (see eq. (3.1.4)), as is schematically illustrated in Fig. 3.3.3. Pictorially, the eventual "contact" between adjacent sheets of the folded surface confers some effective degree of "thickness" to the surface, so that, in the example in Fig. 3.3.3, its effective dimension can be in some sense larger than 2 (and lower than 3). This value is the fractal or Hausdorff dimension.

An important consequence of this effective degree of thickness is that it allows more complex dynamic behaviors; for instance, in Fig. 2.1.4, a trajectory lying on the attractor can repeatedly jump from one lobe to the other without intersecting.

An attractor is called "chaotic" or, equivalently, a system is behaving "chaotically", when it has a sensitive dependence on the initial conditions. This means that two orbits that initially are very close to each other exponentially separate; for long periods both remain within the attractor, but their initial proximity has been completely lost. The exponential separation takes place in the directions of stretching. It is said that any

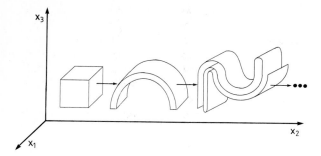

Fig. 3.3.3. Volume contraction in phase space, as a result of the dynamic evolution. There are stretchings and foldings in some directions.

orbit in an attractor is "unstable within the attractor" (i.e. it is stable in the sense that it remains forever on the attractor). Figure 2.1.7 illustrates this fact for the chaotic Lorenz attractor.

Here we are referring to "deterministic" chaos, i.e. to that appearing in systems with a finite number of degrees of freedom or variables which are ruled by a finite set of differential equations, in such a way that if one knows exactly the initial conditions, the future evolution can be exactly predicted. The point is that initial conditions can never be exactly known or repeated, so that the sensitive dependence on these conditions plays a basic role in the dynamic behavior of the system. It entails a loss of predictability over long periods.

The fractal character is more a geometrical than a dynamical property of an attractor, whereas the contrary holds for the chaotic character. All the chaotic attractors found up to now are fractal, but the contrary is not always true. For instance, the aperiodic attractor appearing in the Lorenz model for $r = r_\infty \simeq 139.6$ (Section 3.1.4) is fractal but not chaotic.

The invariance of an attractor under the dynamical evolution entails in general a phenomenon of "self-similarity": an enlarged view of a small zone of the attractor is similar to a larger zone of it. We saw this phenomenon in Fig. 2.2.1 corresponding to the logistic map.

When a dynamical system is described through a map instead of a set of differential equations, it must be taken into account that even for 1-D phase spaces, i.e. for systems described by a unique variable, chaotic and fractal behaviors can be found, as for example in the case of the logistic map shown in Section 2.2. At first sight this fact may seem surprising, but it may be accepted if it is taken into account that in some cases a 1-D map can be obtained from a set of three or more coupled differential equations through a reduction technique (see Section 3.1.2). Indeed, maps generate discrete points which are not affected by the restrictions imposed by the continuity of the solutions of differential equations.

3.3.3 Other related Concepts

a. Non-Wandering Set

From the physical point of view, attractors constitute the most relevant kind of "invariant" sets, i.e. a set of points in the phase space which remain, with evolution of time, within the set. In an actual experiment, almost all the trajectories will asymptotically enter an attractor. From the mathematical point of view, however, trajectories can exist that remain within a bounded region of the phase space without settling on an attractor. They define other kinds of points or sets, which also are physically relevant, because they strongly influence the dynamic behavior of all the nearby trajectories. We briefly describe some of them here.

A point is said to be "non-wandering" if the trajectory originating on it returns to it some time later (or, at least, some of the trajectories originating at arbitrarily close points return to the neighborhood of the points at some later time). The set of all the non-wandering points of a system is called the "non-wandering set".

In addition to attractors, other elements of the non-wandering set are the unstable fixed points and the unstable closed orbits (Section 3.2.1). In spite of the fact that in an actual experiment noise prevents the system from staying at these points or orbits, they play a physical role, because the trajectories can remain close to them for long periods, or they may act as repellors, or may define the border between two basins of attraction, or may become stable under a "structural bifurcation" (Section 3.4), etc.

Other non-wandering subsets are the "attracting sets" (which are defined by only the two first conditions, (i) and (ii), given in Section 3.3.1 for the definition of an attractor) and the homoclinic orbits and cycles and heteroclinic orbits, which are defined below.

b. Homoclinic and Heteroclinic Orbits and Cycles

A "homoclinic orbit" is one asymptotically approaching, for both $t \rightarrow \infty$ and $t \rightarrow -\infty$, the same fixed point, which must therefore be a saddle point, as in the two-dimensional example in Fig. 3.3.4.a. When an orbit asymptotically approaches different fixed points for $t \rightarrow \infty$ and $t \rightarrow -\infty$ it is called a "heteroclinic orbit" (Fig. 3.3.4.b). In fact, in Fig. 3.3.4.b there are two heteroclinic orbits connecting the saddle points P_1 and P_2, so that they define a "homoclinic cycle".

In three-dimensional spaces homoclinic and heteroclinic orbits can be more complex. In Fig. 2.1.2.a a simple homoclinic orbit (dashed line) corresponding to the Lorenz model is shown, and Fig. 3.3.5 shows schematically a Shilnikov homoclinic orbit, which is connected to a spiral saddle point (or "saddle focus") and, as described in subsequent chapters, has recently been observed in laser dynamics.

The dynamics near a homoclinic orbit are very rich, giving rise to "global bifurcations", which are described in Section 3.4.1.

c. Generalized Multistability and Hyperchaos

As is well known, when for a physical system two or more stationary solutions coexist, one speaks about bistability or, in general, multistability. This can be generalized to

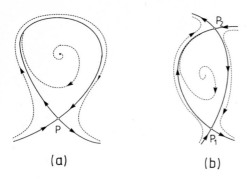

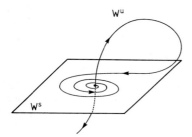

Fig. 3.3.4. (a) A homoclinic orbit (continuous line) connecting a saddle point P to itself. (b) Two heteroclinic orbits (continuous line) defining a homoclinic cycle connecting saddle points P_1 and P_2. Dashed line: other close trajectories.

Fig. 3.3.5. Shilnikov homoclinic orbit of a spiral saddle point (or saddle focus) in a three-dimensional phase space.

the case of solutions settling on any kind of attractor: when two or more attractors coexist in the phase space, one can speak about "generalized multistability". It also entails, as the normal multistability, hysteresis phenomena when some control parameter is continuously varied. The particular case of coexistence of stable periodic solutions is also referred to as "multi-rythmicity". Sudden large perturbations or "hard-mode excitations" of the system may induce it to jump from one basin of attraction to another.

When the boundary between two attractors shows many strong foldings or is fractal, and some degree of noise is present, a trajectory approaching the boundary may jump back and forth across it in an erratic way for a long period. A random sequence of jumps from one to another basin of attraction can then superpose on the chaotic dynamic behavior associated with each attractor, giving rise to a generalized chaos known as "hyperchaos" (Arecchi and Califano, 1986). It can be identified through the Fourier spectrum of the time evolution, which shows a large contribution of the low frequencies with a $\sim 1/f$ dependence, because the time evolution slows down each time the trajectory approaches the fractal basin boundary.

It must be noted that the term "hyperchaos" is also used to denote the chaos appearing when two Lyapunov exponents, instead of only one, become positive.

3.4   Structural Stability: Bifurcations, Roads to Chaos

3.4.1   Bifurcations

In Section 3.2 we have treated the effect on a trajectory of small changes in the initial conditions. Here we are interested in the effect on the trajectories of small changes in the control parameters of the system.

In general, a small variation in one or several control parameters also produces small (continuous) changes in the position and shape of all the attractors of the system in the phase space. A one-to-one mapping between each possible trajectory before and after the small variation can be established, so that the system is said to be "structurally stable". For some specific parameter values, however, one of the attractors or solutions may suffer a strong qualitative change which prevents such a one-to-one mapping. An example is when a fixed point transforms into two close fixed points. This is called a "bifurcation" and the system is said to be "structurally unstable" for this parameter value.

Let us consider here the different kinds of bifurcations that appear more usually .

Local Codimension-One Bifurcations

A bifurcation is said to be "local" when the qualitative changes affecting the bifurcating solution can be analysed by studying only the region of the phase space close to the solution. Also, a bifurcation is said to be of "codimension one" when it satisfies the two following conditions: (i) it can be found by varying only one (any one) of the control parameters of the system, and (ii) a continuous change, within a finite range, in any one of the remaining control parameters does not cause the bifurcation to disappear, but produces only smooth quantitative changes on the bifurcation features*.

Let us consider the local codimension-one bifurcations affecting fixed points and closed orbits.

a. Case of a fixed point

A means of finding the possible bifurcation points affecting a fixed-point solution is to study its dynamic stability. If, when calculating the characteristic exponents (Sections 3.2.2.a and d) as a function of a control parameter μ, one finds that for some value of μ the real part of one of the exponents (or of a pair of complex conjugate exponents) changes from negative to positive values, it means that the stable fixed point becomes unstable and a new kind of stable solution probably sets in. By means of a suitable change of variables, this kind of bifurcation can be described through only one real

---

* In other terms, and in general, a codimension-m bifurcation ($m = 1,2,...$) is that appearing on a set of points of the n-dimensional control parameter space which constitute a manifold of dimension $n - m$ (Bergé, Pomeau and Vidal, 1984). For a study of codimension-two bifurcations, see Guckenheimer and Holmes (1983).

(or, in some cases, complex) variable, which we denote as x; when this is done the problem is said to be reduced to the "normal form" (in general, the normal variable x corresponds to the direction of the eigenvector associated with the characteristic exponent whose real part changes its sign).

Taking as zero the value of the control parameter μ at which the bifurcation appears, and restricting to non-linear terms of order up to 3, the simplest types of bifurcations are as schematically described in Table 3.4.1, where the bifurcation name, the function F appearing in the differential equation $\dot{x} = F(x)$ (see eq. (3.1.2)), and the "bifurcation diagram" showing the stable and unstable branches of fixed-point solutions in the (μ, x) plane are represented. In the case of the Hopf bifurcation, the function F includes an arbitrary constant c which does not play the role of a control parameter, and the variable is complex, $x = x_1 + ix_2$, so that in fact two real variables x_1 and x_2 are needed.

Table 3.4.1 shows only the "supercritical" or "normal" bifurcations, in which the non-linear term contribution is opposite to that of the constant or linear term. By changing the sign of the non-linear term in the function F of Table 3.4.1 the "subcritical" or "inverse" bifurcations are found, the bifurcation diagrams of which are obtained by

Table 3.4.1. Supercritical codimension-one local bifurcations of fixed-points for a system $\dot{x} = F(x)$. The stable (continuous line) and unstable (dashed line) solution branches are shown (in the Hopf case some trajectories eventually approaching the stable solutions have also been depicted).

Name	F	Bifurcation diagram
Saddle–node (or tangent)	$\mu - x^2$	
Transcritical (or with stability exchange)	$\mu x - x^2$	
Pitchfork	$\mu x - x^3$	
Hopf	$(\mu + ic)x - x|x|^2$ $(x = x_1 + ix_2)$	

applying a transformation $(\mu, x) \rightarrow (-\mu, -x)$ to those in Table 3.4.1 and (for the pitchfork and Hopf cases only) by changing the stability character of all the branches (stable $\rightleftarrows$ unstable). In the Lorenz model the fixed points C_+ and C_- become unstable at $r = r_H$ (Section 2.1.3.b) through a subcritical Hopf bifurcation.

It is worth noting that in the subcritical pitchfork (or Hopf) bifurcation the curved unstable branch may fold at a value $\mu' < 0$ (Fig. 3.4.1) and become stable, so that when μ is varied sudden jumps between the stable branches B_0 and B_1 at $\mu = \mu'$ and $\mu = 0$ occur, which give rise to a hysteresis loop (in fact this represents a particular case of generalized multistability, which has been described in Section 3.3.3.c).

b. Case of a Closed Orbit

The bifurcation diagram of a periodic orbit can be studied through the Poincaré map method, using the Floquet theory for the analysis of the surface of section (see Section 3.2.2.c). In this way the problem is reduced to a form similar to that of the previous case, but some differences arise.

The fact that at a bifurcation point the periodic orbit becomes unstable implies that a Floquet multiplier crosses the unit circle in the complex plane (Section 3.2.2.a), which can take place in three ways:

(i) *Through the point $+1$.* In this case the results are very similar to the previous ones, allowing saddle-node, transcritical and pitchfork bifurcations. The function f defining the associated Poincaré return maps $x \rightarrow f(x)$ (or, equivalently, $x_{m+1} = f(x_m)$ as in eq. (3.1.5)) coincides with the corresponding function F in Table 3.4.1, except for the addition of a term x. For instance, for a saddle-node it is

$$f(x) = x + \mu - x^2. \tag{3.4.1}$$

(ii) *Through the point -1.* For this situation, no analog with the fixed-point case exists. A "period doubling" or "subharmonic" or "flip" bifurcation appears, the bifurcation diagram of which looks identical with that of the pitchfork bifurcation of a fixed point, but the dynamics is different: the two parabolic branches do not correspond to independent solutions; instead, there is a unique solution whose return map alternates from one branch to the other indefinitely. In other words, the orbit closes at the second return instead of the first, so that its period becomes twice the original one, as in the example in Fig. 3.4.2. The normal form for the Poincaré map is

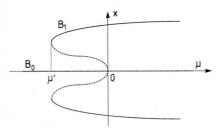

Fig. 3.4.1. A subcritical pitchfork bifurcation at $\mu = 0$ with a branch bending at μ'. Continuous line, stable branch; dashed line, unstable branch.

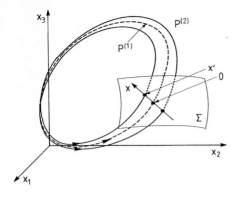

Fig. 3.4.2. An example of period-doubling bifurcation. A closed orbit $P^{(1)}$ of period T (dashed line), crossing the surface of section (Poincaré map) Σ at $x = 0$, transforms into a closed orbit $P^{(2)}$ of period $2\,T$ (continuous line), crossing alternatively at $x = x'$ and $x = -x'$.

$$f(x) = (-1 + \mu)x - x^3 \, .$$

In cases (i) and (ii) above, the variable x corresponds to the direction of the eigenvector associated with the Floquet multiplier $+1$ and -1, respectively.

(iii) *Through a point $\xi \neq \pm 1$.* In this case two complex conjugate Floquet multipliers, ξ and ξ^*, cross the unit circle. The bifurcation is similar to a Hopf bifurcation and it is often known under the same name, but in fact there is one important difference: instead of the circumference (corresponding to a limit cycle) appearing in the bifurcation diagram for $\mu > 0$ in Table 3.4.1, the Poincaré map now gives a series of points also located on a circumference but the order of appearance is such that the angular distance from each point to the consecutive one is constant:

$$x \to f(x) = x\,\mathrm{e}^{i2\pi\theta}, \quad 0 \leqslant \theta < 1 \, . \tag{3.4.2}$$

Thus, if θ is irrational, the whole circumference is covered, denoting quasi-periodic behavior (the limit cycle has bifurcated to a toroidal surface T^2), but if θ is rational only a finite subset of points appears, which corresponds to a closed orbit on the toroidal surface (see Section 3.3.2.b). For $\theta = +1$ and -1, the previous cases (i) and (ii), respectively, are encountered.

Through a similar scheme, the torus T^2 could in turn bifurcate, for another value of the control parameter μ, to an hyper-torus T^3, and so on.

Global (homoclinic) bifurcations

A bifurcation is said to be "global" when the qualitative changes affecting the trajectories in the phase space cannot be deduced from a local analysis in the close neighborhood of a given solution, and a larger zone of the phase space has to be inspected.

A relevant class of global bifurcations is that involving the presence of a homoclinic or heteroclinic orbit (Section 3.3.3.b). Effectively, as Fig. 2.1.2.a (corresponding to the Lorenz-Haken model) shows, the appearance of a homoclinic orbit in general entails a considerable change in the distribution of the dynamical trajectories in the phase space: for values of the control parameter r just below that for which the homoclinic orbit appears (i.e. for $r = 4.9$ in Fig. 2.1.2.a), the global unstable manifold of the origin ends at the point C_+ (see the trajectory in Fig. 2.1.1.a), whereas for values just above ($r > 4.9$) it ends at the point C_-.

If, when continuously increasing a control parameter μ, a homoclinic orbit appears at some value μ_0, immediately afterwards (i.e. for $\mu > \mu_0$) one or many closed orbits are created in the phase space. This can be understood by looking at the simple two-dimensional example in Fig. 3.4.3, where existence, continuity and uniqueness properties of the solutions of differential equations such as eq. (3.1.2) require the appearance of a closed orbit in the region within the loop for $\mu > \mu_0$ (Guckenheimer and Holmes, 1983).

The number of closed orbits created just after a homoclinic bifurcation can be very large. For instance, in the Lorenz model a "strange invariant set" constituted by an infinite number of unstable closed orbits arises just after the appearance of the homoclinic orbit of Fig. 2.1.2 ("homoclinic explosion"). At the beginning these orbits are close together and close to the homoclinic orbit, but as r increases further they move in the phase space and, together with the closed orbits that are being created at successive homoclinic bifurcations (i.e. each time the unstable manifold of the origin returns, after an increasing number of alternative spirallings around the fixed points C_+ and C_- (Fig. 2.1.3), to the origin), they give rise, for $r \geqslant r_A = 35.850$, to the chaotic attractor of Fig. 2.1.4 (Guckenheimer and Holmes, 1983; Sparrow, 1982).

Another basic example is that of a Shilnikov homoclinic orbit (Section 3.3.3.b), when the real positive characteristic exponent (Section 3.2.2.a) of the saddle focus (i.e. that corresponding to the unstable direction) is larger than the modulus of the real negative part of the pair of complex characteristic exponents (i.e. those ruling the attractive spiralling). Under these conditions chaotic behavior occurs close to the Shilnikov orbit, which can be observed, for instance, in the return maps defined near this orbit (Guckenheimer and Holmes, 1983), especially in the sequence of return times, where the chaotic contribution can be observed in general with larger contrast (see, for instance, Section 7.2 and the paper by Arecchi, Meucci and Gadomski (1987) therein cited).

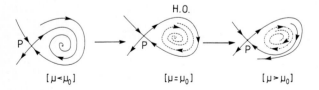

H.O.

$[\mu<\mu_0]$ $[\mu=\mu_0]$ $[\mu>\mu_0]$

Fig. 3.4.3. A simple homoclinic bifurcation appearing at $\mu = \mu_0$. The homoclinic orbit (H.O.) is based on the saddle point P. Continuous line, global stable and unstable manifolds of P; dashed line, other nearby trajectories; dotted-dashed line, a closed orbit arising for $\mu > \mu_0$.

Crises

When a control parameter is varied and two solutions or attractors coalesce, a "crisis" can occur, which consists in a sudden expansion, contraction or disappearance of the attractors. A typical example is a coalescence between a chaotic attractor and an unstable fixed point or periodic orbit, which leads to sudden changes in the chaotic attractor. This occurs, for instance, in the logistic map, on one side of the period-three window (Schuster, 1984; Ott, 1981).

3.4.2   Roads to Chaos: Scenarios

Very often when a control parameter μ is varied and a bifurcation appears at a critical value μ_1, it is followed by a sequence of new bifurcations at higher values μ_i ($i = 2,3,...$), with $\mu_{i+1} > \mu_i$. Each new attractor appearing in the bifurcation chain is usually more complicated than the previous one and eventually it becomes chaotic. The sequence is called a "road (or route) to chaos". The number of types of routes to chaos is unknown, but it has been observed that some appear very often, and for this reason they are called "scenarios". We give here a brief description of the more important scenarios currently known.

a. Ruelle-Takens-Newhouse Scenario

This scenario consists of a sequence of three Hopf bifurcations at critical values μ_1, μ_2, μ_3. As the diagram (3.4.3) shows, the attractor is a fixed point (FP) for $\mu < \mu_1$; at $\mu = \mu_1$ it transforms into a periodic orbit (T); at $\mu = \mu_2$ it changes again to a torus T^2 which entails a quasi-periodic behavior with two incommensurate frequencies, and at $\mu = \mu_3$ a new independent frequency appears, so that in principle a T^3 attractor (hypertorus) would be expected, but in many cases it is unstable towards some kinds of fluctuations and becomes chaotic (CH).

$$(3.4.3)$$

The Ruelle-Takens-Newhouse scenario can be identified by observing the power spectrum $P(f)$ of a dynamic variable, which is similar to the schematic example in Fig. 3.4.4. The broad-band spectrum in (c) corresponds to the chaotic dynamic evolution; some peaks are usually still apparent, which indicate that the previous periodic evolution has not completely disappeared.

As we indicated in Chapter 1, the Ruelle-Takens-Newhouse scenario was the first to be described (1971, 1978) and constituted the first demonstration of the fact that a chaotic behavior can be found in a system with only three degrees of freedom (deterministic chaos). Previously, Landau had supposed that chaos could only appear

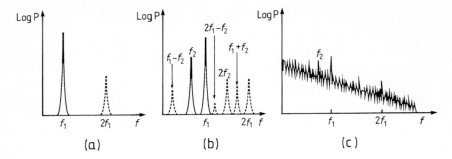

Fig. 3.4.4. Schematic representation of the power spectrum $P(f)$ of a dynamical variable for the three steps of the Ruelle-Takens-Newhouse scenario (see (3.4.3)). (a) Attractor T, with one temporal frequency f_1 (continuous line). (b) Attractor T^2 with frequencies f_1 and f_2. (c) Chaotic attractor, with a broad-band spectrum. If the periodic dynamical evolution is not exactly sinusoidal, then some harmonics and mixing frequencies (dashed lines) appear.

through an infinite sequence of Hopf bifurcations, which can only occur in an infinite-dimensional system.

Curry and Yorke (see Bergé et al., 1984) have demonstrated the existence of a different route for obtaining chaos when, in a sequence like (3.4.3) the torus T^2 has been reached (Bergé, Pomeau and Vidal, 1984). Chaos results from the appearance, when μ is further increased beyond μ_2, of large "wrinkles" on the closed curve defined by the Poincaré map, which folds upon itself, transforming the curve into a fractal object of dimension larger than one, which represents a chaotic attractor.

b. Feigenbaum Scenario (Period Doubling)

Very often a sequence of subharmonic bifurcations (Section (3.4.1.b)) appears in the following way:

$$
\begin{array}{ccccccc}
T & 2T & 4T & 8T & \cdots & & \\
\mid & \mid & \mid & \mid & \text{\small{|||}} \text{-----} \rightarrow & \\
\mu_1 & \mu_2 & \mu_3 & \mu_4 & \mu_\infty & \mu &
\end{array}
\qquad (3.4.4)
$$

where $2^m T$ ($m = 0,1,2,...$) represents the period of a closed orbit attractor (T varies smoothly with μ) and μ_n ($n = 1,2,3,...$) denote the critical values for the parameter μ. Some of the first steps in the cascade might be missing. This sequence is known as the Feigenbaum scenario, which shows the following properties (Eckmann 1981):

(i) $\lim\limits_{m \to \infty} \mu_m = \mu_\infty$ exists.

(ii) At $\mu = \mu_\infty$ the attractor is "aperiodic" (stable periodic orbit of "period 2^∞") and fractal, but not chaotic (see Section 3.3.2.c).

(iii) The difference between consecutive critical values is

$$\lim_{m \to \infty} \frac{\mu_m - \mu_{m-1}}{\mu_{m+1} - \mu_m} = \delta = 4.669201609\ldots \tag{3.4.5}$$

where δ is a universal number.

(iv) In the case of very strong friction, if the critical values μ_1 and μ_2 corresponding respectively to the bifurcations $T \to 2T$ and $2T \to 4T$ exist, then a stable period $3T$ with a large basin of attraction near

$$\mu = \frac{\delta \mu_2 - \mu_1 - \delta(\mu_1 - \mu_2)0.803}{\delta - 1} \tag{3.4.6}$$

can be expected.

(v) After μ_∞ one expects an inverse cascade of "noisy periods":

$$\tag{3.4.7}$$

$(2^m T)$ $(m = 0,1,2,\ldots)$ now represents an attractor yielding a dynamic behavior which is only approximately periodic with mean period $2^m T$. This behavior results from the superposition of the periodic evolution which appeared in the last steps of the direct cascade (i.e. when $\mu \to \mu_\infty$) with an erratic (chaotic) evolution which appears at $\mu = \mu_\infty$ and progressively increases with increasing μ. This chaotic component first only affects (eliminates) the higher harmonics of the periodic evolution (i.e. the components of period $2^m T$, with large m), but as μ increases it progressively affects lower harmonics, giving rise in this way to the inverse cascade of noisy periods.

(vi) Narrow "windows" corresponding to periodic attractors (for instance, of periods $3T$, $5T$, $6T,\ldots$) can be found inside the chaotic domain reached for $\mu \geqslant \tilde{\mu}_1$ in (3.4.7).

(vii) When m steps of the direct cascase (3.4.4) have been observed at the bifurcation points $\mu_1,\ldots,\mu_m$, then it is very probable that a further step will be found at $\mu_{m+1} \simeq \mu_m + (\mu_m - \mu_{m-1})/\delta$. This probability increases with increasing m.

(viii) The mean of the squares of the new amplitudes arising at each successive bifurcation increases until it reaches a level 13.5 dB below that of its predecessors.

Several detailed examples of Feigenbaum scenarios were described in Chapter 2. In Section 2.1.4 (corresponding to the Lorenz model), Figs. 2.1.9, 2.1.10 and 2.1.11 show the closed orbit attractors of periods T, $2T$ and $4T$, respectively, corresponding to the direct cascade (3.4.4); Fig. 2.1.12 shows the attractor of noisy period ($2T$) corresponding to the inverse cascade (3.4.7), and Fig. 2.1.8 shows the attractor corresponding to a periodic window, Fig. 2.2.1, corresponding to the logistic map, also shows period-doubling sequences on different scales, with their direct and inverse cascades, the origin of which is explained in Section 2.2.3.

Often Feigenbaum scenarios are identified by looking at the power spectra characterizing the time evolution of a particular dynamic variable. For instance, Fig. 3.4.5 shows the power spectra corresponding to Figs. 2.1.9, 2.1.10 and 2.1.11 of the Lorenz model. Period doublings are clearly identified in Figs. 3.4.5.b and c through the appearance of subharmonics of the basic frequency $\bar{f} = 1/T$. The power spectrum corresponding to Fig. 2.1.12 would appear to be similar to that of Fig. 3.4.5.b but with a superimposed broad-band spectrum due to the contribution of the chaotic component.

Mathematical similarities between the period-doubling road to chaos and second-order phase transitions can be found in Schuster, 1988; Ott, 1981.

c. Pomeau-Manneville Scenario (Intermittency)

This scenario is also known as the "intermittency" route to chaos from a periodic solution, because it is characterized by the fact that when the control parameter μ exceeds a critical value μ_c, the regular oscillations of the dynamic variable ("laminar" phase) appear to be interrupted at random times by bursts of irregular behaviour ("turbulent" or chaotic phases) (Fig. 3.4.6). The duration of the turbulent phases is fairly regular and weakly dependent on μ, but the mean duration of the laminar phases decreases as μ increases beyond μ_c, and eventually they disappear. Hence only one bifurcation point is associated with the intermittency route to chaos.

Three types of intermittency are known, which depend on the way one of the Floquet multipliers associated with the Poincaré map (Sections 3.4.1.b and 3.2.2.c) leaves the unit circle: through the point $+1$ (type I), -1 (type III) or any other (type II).

The type I intermittency route to chaos begins at a critical value μ_c where a (super- or sub-critical) saddle-node bifurcation appears. The mean duration of the laminar phases increases as $|\mu - \mu_c|^{-\frac{1}{2}}$. A simple model based on a local analysis allows one to understand how intermittency appears: the Poincaré return map describing, for instance, a subcritical saddle-node bifurcation is (see (3.4.1))

$$f(x) = x + (\mu + \mu_c) + x^2 \tag{3.4.8}$$

which has been represented in Fig. 3.4.7 (first row), for $\mu < \mu_c$ and $\mu > \mu_c$.

For $\mu < \mu_c$ the stable (x_-) and unstable (x_+) fixed points associated with the saddle-node bifurcation are clearly identified, but for $\mu > \mu_c$ they have disappeared. In the neck zone ($\mu > \mu_c$ case), however, the return point $f(x)$ remains very close to x, so that the dynamic evolution is almost periodic and gives rise to the laminar phase. From the neck region the periodic behavior quickly disappears and the time evolution is chaotic (turbulent phase). If the map structure outside the neck region allows return through the opposite end ("relaminarization" process), which often occurs, then the periodic behavior reappears again, explaining in this way the intermittency behavior. The statistical distribution of the durations of the laminar phases is given in Bergé, Pomeau and Vidal, 1984; Schuster, 1988; Ott, 1981.

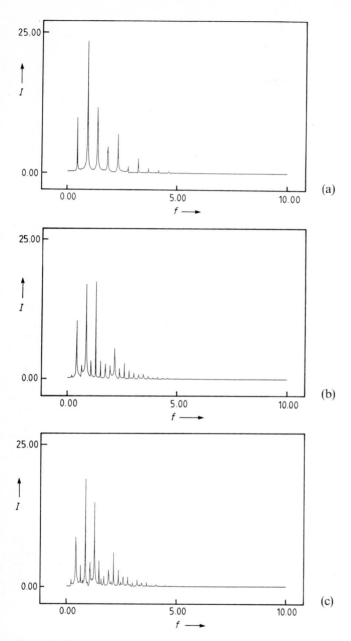

Fig. 3.4.5. Power spectra $I(f)$ corresponding to the evolutions with time of periods (a) T, (b) $2T$ and (c) $4T$ shown, respectively, in Figs. 2.1.9, 2.1.10 and 2.1.11 (Lorenz model). They were obtained by calculating the Fourier transform of $x^2(t)$. The fundamental frequency is $\bar{f} = 1/T$. In (b) and (c) subharmonics at half the frequencies denote the appearance of a period-doubling bifurcation.

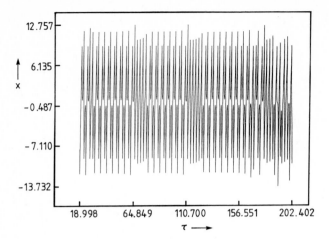

Fig. 3.4.6. Type I intermittency behavior. Laminar phases of almost periodic behavior alternate at random times with turbulent or chaotic phases. This example corresponds to the Lorenz model (Chapter 2), for $r = 118.15$.

This scenario does not allow one to predict exactly when the chaotic regime is reached or what its nature is. It does not have any clear-cut precursors, because the unstable fixed point of the saddle-point bifurcation may not be visible. Increasingly long transients (before the two fixed points collide), or sometimes a cascade of inverse period-doubling bifurcations, might be considered as probable precursors. Time domain measurements are more suitable than power spectra for detecting the scenario.

The type I intermittency road to chaos is found, for instance, close to the windows existing in the chaotic domain, both in the Lorenz model (Section 2.1.3) and in the logistic map (Section 2.2.2). Indeed, each window begins and ends by a road to chaos: by varying μ from the center of a window in one direction or in the other, one obtains chaos through, respectively, an intermittency road or a period-doubling sequence*. Figure 3.4.6, for instance, corresponds to the end of the periodic window reported in Section 2.1.3.c and in Fig. 2.1.8. These roads occur over such a narrow domain, however, that in many cases they cannot be detected.

Type II and III intermittencies have been found less often than type I. They appear at critical values μ_c where a sub-critical bifurcation occurs (Section 3.4.1.b). Their corresponding Poincaré return maps, which are represented in Fig. 3.4.7 (second and third rows), are given by

$$\left. \begin{aligned} \varrho &\to f(\varrho) = \varrho + (\mu - \mu_c)\varrho + b\varrho^3 \\ \theta &\to f(\theta) = \theta + \varphi \end{aligned} \right\} \text{type II} \tag{3.4.9}$$

$$x \to f(x) = -x - (\mu - \mu_c)x - bx^3 \quad \textit{type III}$$

---

* The period-doubling sequence can be affected by "crises" (Section 3.4.1) which lead to a sudden appearance of chaos (Schuster, 1988; Ott, 1981).

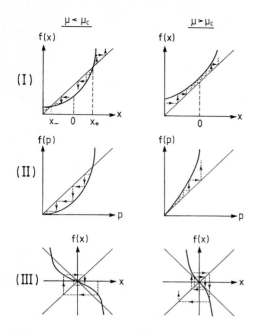

Fig. 3.4.7. Poincaré return maps $x \rightarrow f(x)$ (or, equivalently, $x_{m+1} = f(x_m)$) describing the laminar phases associated, respectively, with type I, II and III intermittency. For $\mu < \mu_c$ (left-hand column) the trajectories (dashed line) converge towards stable fixed points such as x_- (or diverge form unstable fixed points such as x_+), whereas for $\mu > \mu_c$ (right-hand column) the intermittency behavior appears.

For type II intermittency a complex variable $\varrho e^{i\theta}$ has been considered. Because in type III intermittency x is affected by a change of sign, there is an alternation in the peaks appearing when the time evolution of a dynamical variable is recorded: odd-m peaks, for instance, grow with time, whereas even-m peaks decrease with time. This means that the subharmonic amplitude increases and the fundamental amplitude decreases. The chaotic phase appears when the subharmonic amplitude reaches a high value.

Influence of noise

Contrary to older views, it is now clear that external noise is not relevant for the appearance of chaos. Further, small noise influences weakly the form, amplitude and spectrum of the chaos. More precisely, Kifer has demonstrated that all expectation values of bounded observables converge when noise tends to zero (see Eckmann, 1981). In spite of the existence of special ("hyperbolic") points in the attractors the sample paths in the presence of external noise follow, with high probability, some orbit of the deterministic system arbitrarily closely.

The influence of noise on the scenarios can be summarized as follows. The Ruelle-Takens scenario is not destroyed by the addition of small noise, i.e. the chaos is so strong in this case that order cannot be accidentally re-established by small noise. The Feigenbaum cascade is affected by noise on its high periods: to detect one more period doubling, the noise level must diminish by a factor of 6.62, which is very close to the ratio of amplitudes (6.60) between a frequency and its subharmonic. Finally, in the Pomeau-Manneville scenario, the influence of noise is relevant near the bifurcation point μ_c. For instance, for type I intermittency the mean duration of the laminar phases, which is proportional to $(\mu - \mu_c)^{-1/2}$ in the absence of noise, has to be multiplied by $T(\sigma/(\mu - \mu_c)^{3/4})$, where σ is the standard deviation of the noise and T is a universal function (Eckmann, 1981).

The influence of noise in certain cases of generalized multistability has been mentioned in Section 3.3.3.c.

4 Characterization of Chaotic Behavior

To obtain a quantitative and physically meaningful characterization of dynamic behavior is not a simple problem.

The difficulty arises primarily in the case of chaotic motion, because the degree of complexity can be high. One has to quantify in some way "how chaotic" or random a time evolution is and how far in time one can make predictions about the future evolution, or one has to distinguish between deterministic chaos and noise and to determine their relative contributions.

These properties must be described quantitatively by means of appropriate quantities, which have to be calculated from measurements or numerically obtained solutions. The definition of these quantities and a brief discussion of the way in which they are calculated are given in the following sections. We focus attention on asymptotic dynamic behaviors; transients are not considered in this chapter.

4.1 Extraction of Information from Experiments. General Problem

At present, the study of non-linear dynamical systems is performed either experimentally, i.e. through laboratory experiments, or theoretically, i.e. through computer calculations. The kind of information generated is different, so we must consider them separately.

4.1.1 Computer and Laboratory Experiments

Typical features of a "computer experiment" are the following: the differential equations ruling the physical system are known, the time evolution of any variable or observable can be calculated with a relatively high degree of precision and almost any set of initial conditions can be considered. Hence any physical quantity can be calculated through appropriate numerical methods and a detailed characterization of the dynamic behavior can be obtained in principle.

In the case of a laboratory experiment, however, we start with less knowledge: the equations describing the system (taking into account all the experimental factors) are usually unknown; only some of the variables can be measured; only limited numbers of data points can be obtained (time series); and only limited ranges of initial conditions can be probed. This represents a much more limited amount of information from which to characterize the dynamic behavior.

4.1.2   Construction of the Attractor

The first question arising with a laboratory experiment is whether it is possible to obtain enough information about the attractor and its associated dynamics from a time series of data points representing only one (or a few) of the dynamical variables of the system. At present it is believed that this is possible, since the non-linear terms in eqs. (3.1.2) couple all the variables so that all dynamical features appear on all of the variables. Based on this fact, one can construct a multi-dimensional attractor from a time series of measurements corresponding to only one of the system's variables: let us denote by

$$x_i(t_0) , x_i(t_1), \ldots, x_i(t_j), \ldots \qquad (4.1.1)$$

a time series of data corresponding to a dynamical variable x_i, where the time interval $\Delta t = t_j - t_{j-1}$ is assumed to be constant. Now an attractor in a m-dimensional space can be constructed by taking as variables either

$$\vec{x}(t_j) = (x_i(t_j), x_i(t_j + \tau), \ldots, x_i(t_j + (m - 1)\tau)) \qquad (4.1.2)$$

where usually τ is equal to Δt or to a multiple of it; or (if the time derivatives can be obtained with sufficient accuracy),

$$\vec{x}(t_j) = \left(x_i(t_j), \left(\frac{dx_i}{dt} \right)_{t_j}, \ldots, \left(\frac{d^{(m-1)}x_i}{dt^{(m-1)}} \right)_{t_j} \right). \qquad (4.1.3)$$

By plotting $\vec{x}(t_j)$ for each time t_j an attractor different from the "actual" one but topologically equivalent to it results. In other words, we can construct an m-dimensional projection or "image" of the actual attractor (where m is known as the "embedding" dimension) which possesses qualitatively the same dynamical features. It is said that the variables used in eq. (4.1.2) or (4.1.3) are as good for the system's description as the "actual" ones, $x_1(t_j)$, $x_2(t_j)$, ..., $x_i(t_j)$, ...

If one is able to measure more than one variable (say l variables $x_1, x_2, \ldots, x_l$), then l time series such as (4.1.1) will be recorded ($i = 1, 2, \ldots, l$) and the embedding dimension could be increased by a factor l.

Depending on the choice of variables (in particular on the values of m and τ), the attractor's projection will look different. If, for instance, the embedding dimension is smaller than the attractor's Hausdorff dimension, then the projection will be bad (with trajectories that might cross each other). Some theorems state that in general a good projection will be obtained if the number of variables is about twice the Hausdorff dimension (Eckmann and Ruelle, 1985); often, practical criteria based on trial and error are used (see also Section 4.2.2.b).

4.2 Physical Quantities for the Characterization of Dynamic Behavior

To obtain a quantitative and physically meaningful characterization of dynamic behavior, the use of appropriate physical quantities is required. Several of them have already been defined and used for such purposes in the preceding chapters: phase-space trajectories which yield a visual image of the attractor, Poincaré maps, power spectra and Lyapunov exponents. In the first three cases, the way in which these quantities can be obtained from measured or numerically calculated data is simple or well known (for instance, by means of the Fourier transformation for the power spectrum); the case of the Lyapunov exponents is treated in this section.

In addition to the above-mentioned quantities, which directly characterize the time evolution of a system, other quantities have been defined that are based on a statistical analysis of the dynamic motion: generalized dimensions (including Hausdorff, information and correlation dimensions) and entropies, which are related to information theory. They describe the geometrical structure of the attractor (by characterizing the spatial distribution of the points (4.1.2) or (4.1.3) in phase space) and relate it with the dynamical evolution. For this last purpose, the main tool is the ergodic theory, which states that a time average equals a space average. The ergodic theory is usually applied to fluctuating or stochastic systems only. The dynamic behavior described by laws such as (3.1.2) is deterministic, but when the motion lies within a chaotic attractor the loss of long-term predictability brought about by the high sensitivity to initial conditions makes the application of the ergodic theory to such systems also possible (Eckmann and Ruelle, 1985).

4.2.1 Lyapunov Exponents

As defined in Section 3.2.2.d, a Lyapunov exponent λ_i gives a measure of the rate at which the distance $\delta \bar{x}^{(i)}(t)$ between two close solutions increases with time. When $\delta \bar{x}^{(i)}(t)$ remains small, usually one has

$$\delta \vec{x}^{(i)}(t) \simeq \delta \vec{x}^{(i)}(0) e^{\lambda_i t} \tag{4.2.1}$$

Hence the signs of the Lyapunov exponents easily allow one to distinguish between periodic (or quasi-periodic) and chaotic attractors. In the first case there is no exponential separation of initially close trajectories (the distance between two quasi-periodic trajectories is also quasi-periodic) and all the Lyapunov exponents are negative or zero, but the contrary holds in the second case (see Section 3.3.2.d), so that a positive Lyapunov exponent is a clear "signature" of chaos.

Since the Lyapunov exponents are real numbers, it is sufficient for identifying a chaotic attractor to calculate only the largest one. To perform this calculation by using eq. (3.2.11) one is faced with the problem that $\delta \vec{x}^{(i)}(t)$ may become very large and no longer represents an infinitesimal deviation from a given solution $\vec{x}^{(i)}(t)$. Therefore, we describe next some of the methods which have been used or proposed to solve this problem (Eckmann and Ruelle, 1985).

a. Case of a computer experiment

A first method consists in a re-scaling technique. Two close trajectories originating at $\vec{x}(t_0)$ and $\vec{x}(t_0) + \delta \vec{x}(t_0)$ are calculated until $\delta \vec{x}(t)$ becomes fairly large; if this happens at $t = t_1$, then the solution $\vec{x}(t_1) + \delta \vec{x}(t_1)$ is replaced by $\vec{x}(t_1) + \alpha \delta \vec{x}(t_1)$, with $\alpha \ll 1$ (Fig. 4.2.1). In this way, for $t \geqslant t_1$ the trajectories again remain close together, until at $t = t_2$ the re-scaling operation must be repeated, and so on.

Another, better, method makes use of the derivatives of the function F in eq. (3.1.2). By differentiating this equation one obtains

$$\frac{\mathrm{d}}{\mathrm{d}t} \delta \vec{x}(t) = [D_{\vec{x}(t)} F](\delta \vec{x}(t)) \tag{4.2.2}$$

where the $n \times n$ matrix $[D_{\vec{x}(t)} F] (\ldots)$ symbolizes the partial derivatives of $F(\ldots)$ (Jacobian matrix) performed at the point $\vec{x}(t)$. Equation (4.2.2) is linear in $\delta \vec{x}(t)$ with non-constant coefficients. Solving eqns. (3.1.2) and (4.2.2) yields (see eq. (3.1.3))

$$\vec{x}(t) = \mathscr{F}^t(\vec{x}(0)) \quad \text{and} \quad \delta \vec{x}(t) = [D_{\vec{x}(0)} \mathscr{F}^t](\delta \vec{x}(0)) . \tag{4.2.3}$$

The matrices $T_{\vec{x}}^t = D_{\vec{x}} \mathscr{F}^t$ may be calculated by integrating eq. (4.2.2) with n different initial vectors $\vec{u}$, or, better, by solving

$$\frac{\mathrm{d}}{\mathrm{d}t} T_{\vec{x}(0)}^t = [D_{\vec{x}(t)} F] T_{\vec{x}(0)}^t \tag{4.2.4}$$

with $T_{\vec{x}(0)}^0 = \mathbf{1}$ (identity matrix). For computing $T_{\vec{x}}^t$ for large t, it is convenient to discretize time with a small unit τ (not too small, however, because a number of matrices proportional to τ^{-1} have to be multiplied). Next, the calculation proceeds as in the case of a map: $\mathscr{F}^\tau = f$ represents a map in the n-dimensional phase space for which the Jacobian matrix is $D_{\vec{x}} f = T(\vec{x})$, and a product matrix $T_{\vec{x}}^m$ can be defined as

$$T_{\tilde{x}}^m = T(f^{m-1}(\tilde{x})) \dots T(f(\tilde{x})) T(\tilde{x})$$
(4.2.5)

From eq. (3.2.11), it is now evident that the largest Lyapunov exponent is

$$\lambda = \frac{1}{\tau} \lim_{m \to \infty} \left(\frac{1}{m} \right) \log |T_{\tilde{x}}^m (\delta \vec{x})|.$$
(4.2.6)

(For the calculation of the remaining Lyapunov exponents, see Eckmann and Ruelle, 1985).

b. Case of a laboratory experiment

Once the m-dimensional vectors in eq. (4.1.2) have been constructed, one can proceed, as before, in two ways. In one of them an algorithm similar to the first described above is used (Fig. 4.2.1).

Consider the point $\tilde{x}(t_0)$ and its nearest neighbor $\tilde{x}^{(0)}(t_0) = \tilde{x}(t_0) + \delta \tilde{x}^{(0)}(t_0)$. With time the distance between these points increases, until at some time t_1 it exceeds a pre-chosen value ε: $|\delta \tilde{x}^{(0)}(t_1)| \geq \varepsilon$. Substitute now the point $\tilde{x}^{(0)}(t_1)$ by another one $\tilde{x}^{(1)}(t_1)$ $= \tilde{x}(t_1) + \delta \tilde{x}^{(1)}(t_1)$ such that $|\delta \tilde{x}^{(1)}(t_1)| \ll |\delta \tilde{x}^{(0)}(t_1)|$ and $\delta \tilde{x}^{(1)}(t_1)$ is as parallel as possible to $\delta \tilde{x}^{(0)}(t_1)$. From $\tilde{x}(t_1)$ and $\tilde{x}^{(1)}(t_1)$ repeat the same procedure, and so on (the number of points of the original time series must be sufficiently large). The largest Lyapunov exponent can be estimated from the expression

$$\lambda = \frac{1}{(t_{M+1} - t_0)} \sum_{m=0}^{M} \log \frac{\delta \tilde{x}^{(m)}(t_{m+1})}{\delta \tilde{x}^{(m)}(t_m)}$$
(4.2.7)

where the number M is large. More details can be found in the paper by Wolf et al. (1985).

The other method is based, as is the second one above, on the use of the map derivatives $D_{\tilde{x}(t_i)} f_{\tau'} = T_{\tilde{x}(t_i)}^{\tau'}$, where τ' is a multiple of τ and $f_{\tau'}$ is the map which, for $\tilde{x}(t_j)$ close to $\tilde{x}(t_i)$, transforms $\tilde{x}(t_j) - \tilde{x}(t_i)$ to $\tilde{x}(t_j + \tau') - \tilde{x}(t_i + \tau')$. These derivatives are obtained by a best linear fit of $f_{\tau'}$. Once they are known, the calculation proceeds as in the case of a computer experiment. (See Eckmann and Ruelle (1985) for the choice of τ' and other details and comments.)

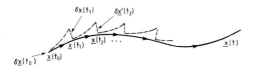

Fig. 4.2.1. Re-scaling method for estimating the largest Lyapunov exponent by numerical calculations (Benettin et al., 1976). Continuous line: a solution $\tilde{x}(t)$ in the phase space. Dashed line: an initially close diverging solution, which is re-scaled at times $t_1, t_2, \dots$

With this method, other positive Lyapunov exponents can also be obtained (the matrix $T_{\tilde{x}(t_i)}^{\tau}$ can only be known with confidence in the unstable directions, for which a larger quantity of experimental points will probably be recorded). For the calculation of the largest Lyapunov exponent, it is not clear which of the two methods has a more general validity.

The dependence of a Lyapunov exponent on a control parameter is not necessarily smooth or continuous. Figure 4.2.2 shows this dependence for the case of the logistic map.

4.2.2   Dimensions

As indicated, dynamic behavior can also be characterized by the geometric or static structure of the associated attractor. This is done by means of a set of related quantities known as "dimensions". The intuitive meaning of the dimension of a set is the number of coordinates or the amount of information needed to specify fully a point on it. The simplest and best known kind of dimension is the Euclidean.

a. Hausdorff Dimension

To define the Hausdorff dimension, let us first introduce the concept of a "covering" (or "coverage") of an object (or set of points) A, which represents a family of sets C_1, C_2, C_3, ... which cover A; the sets C_i ($i = 1,2,...$) may overlap, and they usually consist of arbitrarily small balls or cubes. For computational purposes, the use of cubes might be easier than the use of balls. Obviously, we refer here to cubes with a dimension equal to that of the phase space in which the object A is embedded (see Fig. 4.2.3).

Let us consider a coverage of A by cubes of edge length l, and let $N(l)$ be the minimum number of such cubes required to form the cover. The Hausdorff dimension of A, $D_H(A)$, is defined as

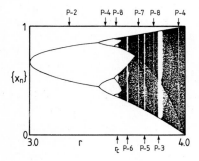
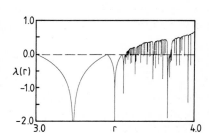

Fig. 4.2.2. a) Iterates of the logistic map, b) Lyapunov exponent λ as a function of r for the logistic map (see Section 2.2).

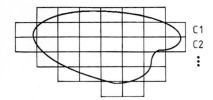

C1
C2
⋮

Fig. 4.2.3. An object A and a covering of A consisting of cubes C_1, C_2, ... of edge length l.

$$D_{\mathrm{H}}(A) = -\lim_{l \to 0} \frac{\log N(l)}{\log l}. \qquad (4.2.8)$$

One may easily prove that for regular objects, such as a point, a segment, a surface or a body in ordinary space, the value of D obtained by eq. (4.2.8) is 0,1,2 and 3, respectively, i.e. the Hausdorff dimension coincides with the geometric or topological dimension and is given by an integer number. In the case of more complicated objects, such as those in Fig. 2.1.6 or the chaotic attractor in Fig. 2.1.4, however, the Hausdorff dimension no longer coincides with the topological one and is not an integer. The objects for which the (non-integer) Hausdorff dimension is larger than the (integer) topological dimension are called "fractal" objects. For instance, for a certain set of parameter values the Hausdorff dimension of the chaotic Lorenz attractor (Fig. 2.1.4) is 2.06. An intuitive interpretation of the Hausdorff dimension was given in Sections 2.1.3.a and 3.3.2.c.

The Hausdorff dimension is also known as the "fractal" dimension. On the other hand, from a strict mathematical point of view, the definition (4.2.8) corresponds to the "capacity" or "Kolmogorov" dimension $D_{\mathrm{K}}(A)$, which fulfils $D_{\mathrm{K}} \geqslant D_{\mathrm{H}}$. For almost all cases, however, $D_{\mathrm{K}} = D_{\mathrm{H}}$ holds.

In the case of a laboratory experiment, the object A can be the set of points $\vec{x}(t_j)$ of eq. (4.1.2) or (4.1.3), which depict an attractor in an m-dimensional phase space. The Hausdorff dimension can be estimated by calculation with eq. (4.2.8). For $D_{\mathrm{H}} > 2$ (and, therefore, $m > 2$), however, this expression is impractical. An alternative procedure consists in calculating other kinds of fractal dimensions, such as those defined below, which represent a lower bound to D_{H} and provide a kind of information equally (if not more) interesting for characterizing the dynamic behavior.

b. Generalized Dimensions. Information and Correlation Dimensions

Let us consider a trajectory of a dynamical system on a strange attractor which is known only at times t_j separated by an interval τ. This is the case, for instance, with the trajectories (4.1.2) or (4.1.3) constructed from a time series of measurements. Let us divide the m-dimensional phase space into boxes of size l^m, and denote by p_i the probability of finding a point $\vec{x}(t_j)$ in the cell number i (or, in other words, the probability with which the system visits the cell number i during the time evolution):

$$P_i = \lim_{N \to \infty} N_i/N \qquad (4.2.9)$$

where N_i is the number of points $\vec{x}(t_j)$ in this cell and N is the total number of available points. An infinite set of "generalized" dimensions D_k ($k = 0,1,2,...$) can then be defined as

$$D_k = \lim_{l \to 0} \frac{1}{k-1} \frac{\log \sum_i p_i^k}{\log l} \qquad (4.2.10)$$

where the sum extends over the total number of cells, $N(l)$.

The D_0 dimension is the Hausdorff dimension (compare eq. (4.2.10) with eq. (4.2.8) for $k = 0$). The D_1 dimension is

$$D_1 = -\lim_{l \to 0} S(l)/\log l \qquad (4.2.11)$$

where

$$S(l) = -\sum_i p_i \log p_i \qquad (4.2.12)$$

is an entropy-like quantity. Since the entropy is a concept related with the information theory (Shannon and Weaver, 1949) (it gives, in our case, a measure of the degree of disorder of the points on the attractor or, in other words, of the amount of information, in "bits", needed to locate the system in a certain cell i), D_1 is known as the "information dimension". It is easy to verify that for an uniform attractor ($p_i = 1/N(l)$) D_0 and D_1 coincide, so that the difference between these dimensions gives a measure of the non-uniformity of the attractor (note that D_0 is independent of the specific spatial distribution of points within the attractor, i.e. it does not depend on the p_i, whereas D_1 takes into account this distribution). The D_2 dimension is given by

$$D_2 = \lim_{l \to 0} \frac{\log \sum_i p_i^2}{\log l}. \qquad (4.2.13)$$

For large N, the sum $\sum p_i^2$ represents the probability that any pair of points lie in the same cell. This approximately coincides with the "correlation integral" $C(l)$, which gives the probability that any pair of points is separated by a distance smaller than l:

$$C(l) = \lim_{N \to \infty} \frac{1}{N^2} \sum_{j,j'} \Theta(l - |\vec{x}_j - \vec{x}_{j'}|) \qquad (4.2.14)$$

where $\vec{x}_j = \vec{x}(t_j)$ and Θ represents the Heaviside function ($\Theta(y) = 0$ for $y < 0$ and $\Theta(y) = 1$ for $y \geqslant 0$). For this reason, D_2 is known as the "correlation dimension". Note that eq. (4.2.14) gives easier numerical calculations from a time series than $\sum_i p_i^2$.

Grassberger and Proccaccia (1983a and b) developed a practical algorithm for computing D_2 based on the fact that $C(l)$ is in practice proportional to l^v for small l, and v almost coincides with D_2. This method also allows one to determine the minimal embedding dimension of the attractor, m_0 (see Section 4.1.2), because for $m \geqslant m_0$ one obtains a v value independent of m (and almost equal to D_2), whereas for $m < m_0$ one obtains a lower value proportional to m (Eckmann and Ruelle, 1985; Bergé, Pomeau and Vidal, 1984). The correlation integral is also useful for distinguishing between deterministic chaos and noise: when the amplitude of the fluctuations in the phase space produced by noise is much smaller than the cell dimension l, then the slope of a plot of $\log C(l)$ versus $\log l$ yields the D_2 value of the attractor (deterministic behavior), but when the noise amplitude is larger than l, this slope yields a larger value equal to the dimension m of the space in which the attractor has been embedded (Schuster, 1984).

The general rule $D_{k'} \leqslant D_k$ for $k' > k$ is fulfilled (in the case of the equality sign, only for uniform attractors). As an example, for the logistic map in Section 2.2 the dimensions D_0, D_1, D_2 and D_∞ are, respectively, 0.538, 0.537, 0.500 and 0.394 for $\mu = \mu^{(\infty)}$, whereas they take the common value 1 for $\mu = 4$.

A mathematical relation between the Lyapunov exponents (i.e. a dynamic property) and the D_0 dimension (i.e. a static property) for an attractor is given by the Kaplan-Yorke conjecture (Schuster, 1988).

4.2.3 Kolmogorov Entropy

Consider, as before, a trajectory such as (4.1.2) or (4.1.3) in an m-dimensional phase space which has been divided into boxes of size l^m. Let $p_{i_0,\ldots,i_n}$ be the joint probability that $\vec{x}(0)$ lies in the cell i_0, $\vec{x}(\tau)$ in the cell $i_1,\ldots$, and $\vec{x}(n\tau)$ in the cell i_n. Then, by comparison with the usual entropy concept in information theory (cf. eq. (4.2.12)), the quantity

$$K_n = -\sum_{i_0,\ldots,i_n} p_{i_0,\ldots,i_n} \cdot \log p_{i_0,\ldots,i_n} \tag{4.2.15}$$

is a measure of the information needed to specify an n-point long trajectory with a precision l. As a consequence, $K_{n+1} - K_n$ measures the loss of information from time $n\tau$ to time $(n + 1)\tau$. The Kolmogorov entropy, or "K-entropy", is defined as the average rate of loss of information:

$$K = \lim_{\tau \to 0} \lim_{l \to 0} \left\{ \lim_{N \to \infty} \frac{1}{N\tau} \sum_{n=0}^{N-1} (K_{n+1} - K_n) \right\}$$

$$= -\lim_{\tau \to 0} \lim_{l \to 0} \left\{ \lim_{N \to \infty} \frac{1}{N\tau} \sum_{i_0,\ldots,i_{N-1}} p_{i_0,\ldots,i_{N-1}} \cdot \log p_{i_0,\ldots,i_{N-1}} \right\} \tag{4.2.16}$$

where the limit $\tau \to 0$ must be omitted in the case of a map.

The Kolmogorov entropy is the fundamental quantity for measuring the "degree of chaoticity" of a particular motion. Its lower value is zero and corresponds to regular time evolution; for chaotic motions it is positive (the more positive, the "more chaotic" is the system) and for perfectly random systems it is infinite. It allows one to define a strange attractor as an attractor with a positive K-entropy.

The Kolmogorov entropy can be related with the Lyapunov exponents (Pesin, 1977)

$$K = \int \varrho(x) \left(\sum_i \lambda_i^+ (\vec{x}) \right) d\vec{x} \tag{4.2.17}$$

where $\varrho(x)$ represents the density of points in the attractor, which is independent of time ("invariant measure"), and λ_i^+ represent the positive Lyapunov exponents. For a one-dimensional map, K equals the (unique) Lyapunov exponent. For higher dimensional phase spaces, K measures the average deformation rate of a cell in phase space with evolution of time or, equivalently, is inversely proportional to the time interval over which the future evolution can be predicted (Schuster, 1984).

Grassberger and Proccaccia (1983c) developed a practical method for obtaining a lower bound to the Kolmogorov entropy (note that the box counting method suggested by eq. (4.2.16) is impractical). This bound is given by

$$K_2 = \lim_{l \to 0} \lim_{n \to \infty} \log \frac{C_n(l)}{C_{n+1}(l)} \tag{4.2.18}$$

where $C_n(l)$ is a generalization of the correlation integral $C(l)$ of eq. (4.2.14) and represents the probability that any pair of points $\vec{x}_j$, $\vec{x}_{j'}$ fulfill

$$\sum_{k=0}^{n-1} (\vec{x}_{j+k} - \vec{x}_{j'+k})^2 < l^2$$

For small l and large embedding dimension m it is found that

$$C(l) \sim l^{D_2} e^{-m \tau K_2} \tag{4.2.19}$$

so that in practice K_2 can be determined by calculating $C(l)$ for different values of m:

$$K_2 = \tau^{-1} [\ln C(l)_m - \ln C(l)_{m+1}] \tag{4.2.20}$$

and taking the asymptotic value for large m (see Schuster (1988), Eckmann and Ruelle (1985) and Grassberger and Proccaccia (1983c) for further details). Because $K_2 \leqslant K$, the condition $K_2 > 0$ is sufficient for the existence of chaos.

4.3   Practical Examples

As described before, for characterizing experimental data an equivalent attractor is constructed by delay of a measured time series with respect to itself. The point set constructed in this way, as indicated, has some important topological properties of the attractor, in particular its dimension. A question is how to choose the delay τ properly.

A rule of thumb states that τ should be the time in which the autocorrelation function of the time series $x(t_1),\ldots,x(t_N)$ has fallen to e^{-1}. A more intuitive way of selecting τ that is largely equivalent to this rule of thumb consists in the following.

Plot values of $x(t_k)$ versus $x(t_{k+\tau})$ for all $k = 1\ldots N$. If the points then lie on a straight line, τ is too short because, evidently, the points are strongly correlated. If the points form a structureless cloud, τ is too long and the points look like random noise (with an infinite dimension). Somewhere in between one can (for a chaotic system) find a plot with a pronounced structure for which the τ is then probably a "good" value.

After having constructed these point sets in the spaces of increasing dimensions very closely following the definition of dimension, for each embedding dimension around certain points, the number of other points within an m-dimensional sphere is counted, and this is done for successive increasing values of the radius l of the sphere.

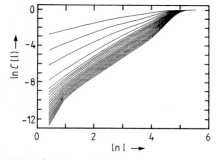

Fig. 4.3.1. Correlation integral $C(l)$ for increasing values of the embedding dimension m (curves from top to bottom: $m = 1,2,3,\ldots$). It corresponds to the pseudo-attractor constructed from the intensities calculated from the Lorenz model (Fig. 4.3.3) for a delay τ of 0.15 periods. To simulate the detector noise of an experiment, Gaussian noise of 0.4% of the maximum intensity at half-width has been added to the intensity.

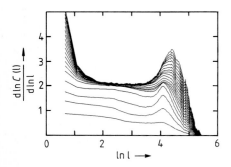

Fig. 4.3.2. Derivative of Fig. 4.3.1. The "plateau" height gives the dimension D_2.

This allows one to obtain the correlation integral $C(l)$, which according to Sect. 4.2.2 is plotted against the radius l of the sphere in a double logarithmic representation to find the exponent of the growth law. If the phase-space dimension m is chosen to be smaller than the actual dimension of the attractor, it cannot properly accommodate the equivalent attractor and the growth exponent will underestimate the attractor dimension. Therefore, this procedure is repeated from $m = 1$ until a value of m where the exponent (or the slope of the $C(l)$ curve in the double logarithmic representation) no longer increases with increasing m.

The final value to which the growth exponent tends with increasing m then gives the attractor dimension D_2. The value m at which the slope reaches its final value is called the "embedding dimension", as it is large enough to accommodate the attractor. Obviously, this embedding dimension is a lower bound of the degrees of freedom of the system and thus of the number of first-order differential equations describing it.

As an example, Fig. 4.3.1 shows the curve $C(l)$ for different values of m, calculated for a solution of the Lorenz model. It can be clearly seen that the slope in the double logarithmic representation tends to a constant value with growing m (embedding dimension). To show this more clearly, the derivatives $d \log C(l)/d \log l$ have been plotted in Fig. 4.3.2.

Figure 4.3.2 shows a clear "plateau" for intermediate values of l towards which all curves converge. To simulate an experimental signal which contains noise, to the calculated trajectories of the Lorenz equations a random noise of Gaussian amplitude distribution with 0.4% Gaussian distribution half-width has been added. This manifests itself in the rising of $d \log C(l)/d \log l$ towards small l. This can be easily understood:

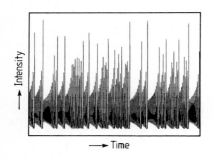

Fig. 4.3.3. Intensity pulses calculated from the Lorenz model with parameters $R = 15$, $b = 0.25$ and $\sigma = 2$, corresponding to the laser experiments (Chapter 5).

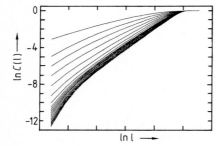

Fig. 4.3.4. The same as in Fig. 4.3.1 but for experimentally measured ammonia laser pulses (Chapter 5).

for small l the noise amplitudes are larger than the differences between vector tips. Noise has no finite dimension, and consequently one "sees" only noise, whose dimension increases without bounds as the embedding dimension. The "hump" at large l is not a general feature but is associated with some macroscopic structure of the attractor. Figure 4.3.3 shows the intensity trajectories of the Lorenz model as used for the dimension calculation. It can be seen that the "attractor" has a "hole", i.e. the trajectories are avoiding the central area (see Fig. 2.1.4). When l reaches a value of roughly the "diameter" of the attractor, the number of points within a sphere of radius l suddenly includes the points on the "opposite side" of the attractor and therefore at that point are growing more rapidly than before, thus producing the "hump" in $d \log C(l)/d \log l$. This hump is an artifact from the mathematical method. According to the proper definition of dimension, the radius l should always be small compared with the attractor. For the experimental problem of never having enough points measured, in experimental dimension determinations l is usually taken as large as possible. This then results in cases like Fig. 4.3.3 in such artifacts. In cases in which the attractor is more uniform and less structured, the "hump" does not occur.

This determination of dimension from the Lorenz attractor can be compared with that from a real experiment. As described in Chapter 5, it is found that an ammonia laser behaves very much like the predictions of the Lorenz model for the homogeneously broadened laser.

Figure 4.3.4 shows the growth curves $\log C(l)$ calculated from the measured intensity pulsing (Fig. 5.2.18a). For a clearer view of the dimension, the slopes in Fig. 4.3.4 are shown in Fig. 4.3.5. The "plateau" clearly gives a dimension of 2.1, consistent with the calculations (Fig. 4.3.2) on the Lorenz model. The noise in the experimental data clearly shows the same effect at low l values as with the artificial noise in the Lorenz model data (Fig. 4.3.2).

It should be mentioned that the RMS noise in the experimental data is fairly small. Although clearly visible in the "slopes" in Fig. 4.3.5, in the experiment the signal-to-noise ratio was 500. This indicates that very low noise measurements are needed for reliable dimension calculations. In any case, the good agreement between the measured values (Fig. 4.3.5) and the values obtained from Lorenz model data shows that fractal dimensions can be utilized to test the adequacy of models for experiments.

An even better test for comparing the experiment with the model is, of course, to determine the generalized dimensions in eq. (4.2.10). Figure 4.3.6 shows the generalized

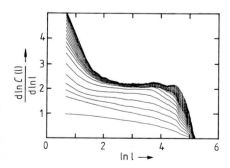

Fig. 4.3.5. Derivative of Fig. 4.3.4. The plateau height gives the dimension D_2.

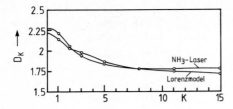

Fig. 4.3.6. Comparison of generalized dimensions $D_0 - D_{15}$ of the Lorenz model calculation (Fig. 4.3.3) and the ammonia laser experiments (Chapter 5).

dimensions for the Lorenz model with parameters as used before and the measurements on the laser as used before. It is seen that the Hausdorff dimension ($k = 0$) is the largest and, as predicted, the dimension decreases with k, clearly showing the non-uniformity of the attractors. Note again the good agreement (within a few percent) between measured laser data and Lorenz model data.

5 Two-Level Single-Mode Gas Lasers

In this and subsequent chapters we discuss the dynamic behavior of the different kinds of lasers. The theoretical understanding of these behaviors is based on the analysis of the set of coupled non-linear differential equations describing each kind of laser. This analysis is performed following the approach described in Chapter 3.

In the first section we introduce the laser equations by briefly recalling how they are obtained from the fundamental equations of electromagnetism and quantum physics. For the sake of clarity we do this for the case of the simplest kind of laser, the "two-level homogeneously broadened laser". In subsequent sections we describe the dynamic behaviors that have been predicted or observed in two-level homogeneously broadened and inhomogeneously broadened lasers.

5.1 Laser Equations

The theoretical description of the radiation amplification process in a laser requires the use of adequate physical variables for characterizing the time evolution of the two subsystems that interact within the resonator: the generated wave and the amplifying medium. A good choice of variables is the electric field of the generated wave and the atomic dipole moment (or the polarization) and the "population inversion" of the atoms or molecules of the amplifying medium.

The equation describing the evolution of the wave electric field $\vec{\mathscr{E}}(\vec{x},t)$ can be obtained from the well known wave equation of electromagnetic theory:

$$\Delta\vec{\mathscr{E}} - \frac{1}{c^2}\ddot{\vec{\mathscr{E}}} - \mu_0\sigma\dot{\vec{\mathscr{E}}} = \mu_0\ddot{\vec{\mathscr{P}}} \qquad (5.1.1)$$

where $\Delta = \partial^2/\partial x^2 + \partial^2/\partial y^2 + \partial^2/\partial z^2$ represents the Laplace operator, c is the velocity of light in vacuum, μ_0 is the vacuum magnetic permeability ($\mu_0\varepsilon_0 = 1/c^2$), σ is the electric conductivity of the medium, $\vec{\mathscr{P}}(\vec{x},t)$ is the electric polarization of the medium and the dot represents the partial derivative with respect to time. To deal with simple expressions we shall restrict ourselves to the case of a laser with a ring resonator in which only one mode is excited (uni-directional single-mode laser). Further, we shall neglect the transverse spatial structure of the field, i.e. we shall assume a plane-wave structure, so that $\vec{\mathscr{E}}(\vec{x},t)$ and $\vec{\mathscr{P}}(\vec{x},t)$ can be expressed as

$$\vec{\mathscr{E}}(\vec{x},t) = \frac{1}{2}[\vec{\mathscr{E}}(t)\vec{e}\,\mathrm{e}^{\mathrm{i}\vec{k}\cdot\vec{x}} + \text{c.c.}]$$

$$\vec{\mathscr{P}}(\vec{x},t) = \frac{1}{2}[\mathscr{P}(t)\vec{e}'\,\mathrm{e}^{\mathrm{i}\vec{k}\cdot\vec{x}} + \text{c.c.}]$$

(5.1.2)

where $\vec{k}$ is the resonator mode wavevector and $\vec{e}$ and $\vec{e}'$ are unit vectors. For simplicity we may assume $\vec{e}'$ to be parallel to $\vec{e}$, since only the component of $\vec{\mathscr{P}}$ parallel to $\vec{e}$ is dynamically coupled to $\vec{\mathscr{E}}$ through eq. (5.1.1). On the other hand, we shall consider media with zero electric conductivity, so that the third term in eq. (5.1.1) describing the field absorption would in principle vanish. However, we shall introduce an "effective" electric conductivity σ which describes the field attenuation brought about by the mirror's transmission, diffraction, diffusion, etc. If the cavity losses are characterized by a decay rate κ, then σ is

$$\sigma = 2\varepsilon_0\kappa.$$

(5.1.3)

Note that in this way the mirror's transmission losses are uniformly distributed along the resonator instead of being concentrated on the mirrors.

To further simplify eq. (5.1.1), we introduce the so-called "slowly varying amplitude approximation" (Sargent, Scully and Lamb, 1974; Haken, 1985), which is based on the fact that the field and polarization amplitudes usually vary little in an optical period, i.e. if, for instance, $\mathscr{E}(t)$ is expressed as

$$\mathscr{E}(t) = A(t)\mathrm{e}^{-\mathrm{i}\omega_c t}$$

(5.1.4)

where $\omega_c = ck$ ($k = 2\pi/\lambda$) is the cavity-mode frequency (in vacuum), then

$$|\dot{A}| \ll |\omega_c A|, \quad |\ddot{A}| \ll |\omega_c \dot{A}|.$$

(5.1.5)

This allows several terms in the wave equation to be neglected, so that with eqns. (5.1.2), (5.1.3) and (5.1.5), eq. (5.1.1) transforms into

$$\dot{\mathscr{E}}(t) = -(\mathrm{i}\omega_c + \kappa)\mathscr{E}(t) + \frac{\mathrm{i}\omega_c}{2\varepsilon_0}\mathscr{P}(t).$$

(5.1.6)

The equations describing the atomic dipole moment (or the polarization) and the population inversion of the atoms of the laser medium can be obtained from the Schrödinger equation:

$$i\hbar \frac{\partial \Psi(t)}{\partial t} = [H_0 + H_{\text{int}}(t)] \, \Psi(t) \qquad\qquad (5.1.7)$$

where $\Psi(t)$ represents the atomic wave function, H_0 the Hamilton operator of the unperturbed atom and $H_{\text{int}}(t)$ the interaction Hamiltonian describing the action of the laser field on the atom.

If we assume that the laser field is resonant or quasi-resonant only with a transition between two levels a and b (Fig. 5.1.1) of energies E_a and E_b, respectively, then we can disregard other atomic levels ("two-level approximation") and express $\Psi(t)$ as

$$\Psi(t) = c_a(t) \, \Psi_a \, \mathrm{e}^{-iE_at/\hbar} + c_b(t) \, \Psi_b \, \mathrm{e}^{-iE_bt/\hbar} \qquad\qquad (5.1.8)$$

where $c_a(t)$ and $c_b(t)$ are time-dependent coefficients and the eigenfunctions Ψ_a and Ψ_b depend on the atomic electron coordinates and satisfy

$$H_0 \Psi_a = E_a \Psi_a, \quad H_0 \Psi_b = E_b \Psi_b. \qquad\qquad (5.1.9)$$

On the other hand, in the dipole electric approximation $H_{\text{int}}(t)$ is given by

$$H_{\text{int}}(t) = -\tilde{\vec{\mu}} \cdot \vec{\mathscr{E}}(\vec{x}_n, t) \qquad\qquad (5.1.10)$$

where $\tilde{\vec{\mu}}$ represents the electric dipole operator and $\vec{x}_n$ denotes the fixed spatial position of the atom that we are considering (let us denote it as the atom n). The field $\vec{\mathscr{E}}$ must be the laser field considered above (except for local field corrections, which we neglect as we are considering dilute media).

Assuming that the atomic states Ψ_a and Ψ_b do not have permanent dipole moments, i.e. $\vec{\mu}_{aa} = \vec{\mu}_{bb} = 0$ (where $\vec{\mu}_{aa} \equiv \langle \Psi_a | \tilde{\vec{\mu}} | \Psi_a \rangle$, etc.), and that $\vec{\mu}_{ba} = \vec{\mu}_{ab}$ is a real vector (it can always be chosen in this way), the introduction of eq. (5.1.8) into eq. (5.1.7), using eqns. (5.1.9) and (5.1.10) leads to the following pair of coupled equations for the coefficients $c_a(t)$ and $c_b(t)$:

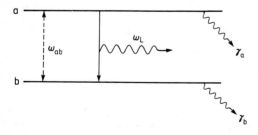

Fig. 5.1.1. Two-level system, consisting of levels a and b separated in energy by $\hbar\omega_{ab}$, coupled with a laser field of angular frequency ω_{L}. γ_a and γ_b are the population relaxation rates of each atomic level.

$$\dot{c}_a(t) = \frac{i}{\hbar} \vec{\mu}_{ab} \cdot \vec{\mathscr{E}}(\vec{x}_n, t) e^{i\omega_{ab}t} c_b(t)$$

$$\dot{c}_b(t) = \frac{i}{\hbar} \vec{\mu}_{ab} \cdot \vec{\mathscr{E}}(\vec{x}_n, t) e^{-i\omega_{ab}t} c_a(t) \tag{5.1.11}$$

where $\omega_{ab} = (E_a - E_b)/\hbar$.

However, instead of solving eqs. (5.1.11) with the variables $c_a(t)$ and $c_b(t)$, which do not have a direct physical meaning, one can try to establish a system of equations for the induced electric dipole and population inversion. Since the induced atomic dipole $\vec{\mu}(t)$ is defined as the expectation value of $\tilde{\vec{\mu}}$:

$$\vec{\mu}(t) = \langle \Psi(t) | \tilde{\vec{\mu}} | \Psi(t) \rangle$$
$$= c_b^*(t) c_a(t) e^{-i\omega_{ab}t} \vec{\mu}_{ba} + \text{c.c.} \tag{5.1.12}$$

let us choose as a dynamical variable the product of coefficients appearing in eq. (5.1.12), which we denote by $p_n(t)$:

$$p_n(t) \equiv c_b^*(t) c_a(t) e^{-i\omega_{ab}t}. \tag{5.1.13}$$

The second dynamical variable is directly the population inversion d_n:

$$d_n \equiv |c_a(t)|^2 - |c_b(t)|^2. \tag{5.1.14}$$

In eqns. (5.1.13) and (5.1.14) the subscript n denotes, as before, that we are dealing with an atom located at $\vec{x}_n$.

By using eqn. (5.1.11), one can easily deduce that $p_n(t)$ and $d_n(t)$ obey the following system of equations:

$$\dot{p}_n(t) = -i\omega_{ab} p_n(t) - \frac{i}{\hbar} \vec{\mu}_{ab} \cdot \vec{\mathscr{E}}(\vec{x}_n, t) d_n(t)$$

$$\dot{d}_n(t) = \frac{2i}{\hbar} \vec{\mu}_{ab} \cdot \vec{\mathscr{E}}(\vec{x}_n, t) [p_n^*(t) - p_n(t)]. \tag{5.1.15}$$

If we introduce in eqn. (5.1.15) the expression (5.1.2) for $\vec{\mathscr{E}}(t)$ and take into account that p_n contains a factor $e^{-i\omega_{ab}t}$ and $d_n(t)$ does not contain any exponential factor of this type, it turns out that the terms containing factors $\mathscr{E}(t)^* d_n(t)$, $\mathscr{E}(t) p_n(t)$ or $\mathscr{E}(t) p_n(t)^*$ can be neglected. In effect, they are multiplied by time-exponential factors oscillating much faster than those of the remaining terms, so that when integrated over a time larger than one optical period their contribution vanishes ("rotating-wave approximation"). Further, if we express $p_n(t)$ and $d_n(t)$ as

$$p_n(t) = p(t)e^{i\vec{k}\cdot\vec{x}_n}$$

$$d_n(t) = d(t)$$

(5.1.16)

then the exponential factor $\exp(i\vec{k}\cdot\vec{x}_n)$ disappears from eqs. (5.1.15), which transform into

$$\dot{p}(t) = -i\omega_{ab}p - \frac{i}{2\hbar}\mu_{ab}^{\parallel}\mathscr{E}(t)d(t)$$

$$\dot{d}(t) = \frac{i}{\hbar}\mu_{ab}^{\parallel}[\mathscr{E}(t)p^*(t) - \mathscr{E}^*(t)p(t)]$$

(5.1.17)

where $\mu_{ab}^{\parallel} = \vec{\mu}_{ab}\cdot\vec{e}$ represents the projection of $\vec{\mu}_{ab}$ on the electric field direction $\vec{e}$. These equations are independent of the atomic index n, i.e. they are sufficient to describe the dynamic evolution of the whole laser medium. Note that $p(t)$ is directly related to the medium polarization $\mathscr{P}(t)$ defined in eq. (5.1.2):

$$\frac{1}{2}\mathscr{P}(t) = \frac{N}{V}p(t)\mu_{ba}^{\parallel}$$

(5.1.18)

where N represents the total number of atoms in the laser medium of volume V.

In the derivation of eqns. (5.1.17), for simplicity we have ignored any factor giving rise to inhomogeneous broadening, such as Doppler effects in a gaseous medium. Homogeneous broadening factors, such as spontaneous emission and collisions, must be taken into account in eqns. (5.1.17). This is usually done through phenomenological transverse, $\gamma_\perp$, and longitudinal, $\gamma_\parallel$, relaxation rates which describe the damping of the dipole moment $p(t)$ and the population inversion $d(t)$, respectively. Hence one has to add a term $-\gamma_\perp p(t)$ to the first of eqns. (5.1.17) and a term $-\gamma_\parallel[d(t) - d_0]$ to the second, where d_0 represents the population inversion in the absence of laser field (i.e. that created by the pumping mechanism).

Writing together eq. (5.1.6) and eqns. (5.1.17), one obtains the final set of laser equations:

$$\dot{E}(t) = -(i\omega_c + \kappa)E(t) + Ngp(t)$$

$$\dot{p}(t) = -(i\omega_{ab} + \gamma_\perp)p(t) + gE(t)d(t)$$

$$\dot{d}(t) = -\gamma_\parallel[d(t) - d_0] - 2g[E(t)p(t)^* + \text{c.c.}]$$

(5.1.19)

where

$$g = \left(\frac{\mu_{ab}^{\parallel 2}\omega_c}{2\hbar\varepsilon_0 V}\right)^{1/2}, \quad E(t) = -i\left(\frac{\varepsilon_0 V}{2\hbar\omega_c}\right)^{1/2}\mathscr{E}(t).$$

(5.1.20)

The change of variable $\mathscr{E}(t) \to E(t)$ has been introduced in order to describe the system with three dimensionless variables $E(t)$, $p(t)$ and $d(t)$. The set of eqns. (5.1.19) is known as the semi-classical Maxwell-Bloch-type equations for a two-level homogeneously broadened single-mode unidirectional laser, in the plane-wave uniform-field approximation.

The first term in the first two eqns. (5.1.19), i.e. $-i\omega_c E(t)$ and $-i\omega_{ab} p(t)$, describes a fast-oscillating factor in $E(t)$ and $p(t)$ of frequency equal (or close) to ω_c and ω_{ab}. The terms containing the factors κ, $\gamma_\perp$ or $\gamma_\parallel$ describe the damping of the corresponding variables, and the last term in each equation describes the coupling between the variables brought about by the radiation-matter interaction.

Equations (5.1.19), or other similar equations for different kinds of lasers, will be used in this and subsequent chapters. In some cases, however, another form of the laser equations will be used (Sargent, Scully and Lamb, 1974), in which the atomic medium is described by the "ensemble-averaged" or "population" density matrix $\varrho(z,t)$ (where z is the coordinate along the field propagation direction), the matrix elements of which describe the level populations and induced polarization created at time t and point z by all the atoms located at this point. Within the semi-classical theory, the density matrix obeys the evolution equation

$$\frac{\partial}{\partial t}\varrho(z,t) = -\frac{i}{\hbar}[H_0 + H_{int}(z,t),\varrho(z,t)] - \Gamma\varrho(z,t) \tag{5.1.21}$$

where the brackets [,] denote the quantum commutation operation and $H_{int}(z,t) = -\vec{\tilde{\mu}} \cdot \vec{\mathscr{E}}(z,t)$.

For a two-level atomic system such as that in Fig. 5.1.1, the matrix ϱ is

$$\varrho = \begin{pmatrix} \varrho_{aa} & \varrho_{ab} \\ \varrho_{ba} & \varrho_{bb} \end{pmatrix} \tag{5.1.22}$$

with $\varrho_{ab} = \varrho_{ba}^*$. If ϱ is normalized, for instance, to one atom, the diagonal elements $\varrho_{aa}(z,t)$ and $\varrho_{bb}(z,t)$ represent the population of, respectively, the levels a and b of an atom located at point z at time t, whereas the non-diagonal element $\varrho_{ab}(z,t)$ is proportional to the medium polarization $\mathscr{P}(z,t)$:

$$\mathscr{P}(z,t) = \frac{N\mu_{ab}}{V}\varrho_{ab}(z,t) + \text{c.c.} \tag{5.1.23}$$

When eq. (5.1.22) is inserted into eq. (5.1.4), the following set of equations results:

$$\dot{\varrho}_{ab} = -(i\omega_{ab} + \gamma_\perp)\varrho_{ab} + i\hbar^{-1}V_{ab}(z,t)(\varrho_{aa} - \varrho_{bb})$$
$$\dot{\varrho}_{aa} = -\gamma_a(\varrho_{aa} - \varrho_{aa}^{(0)}) - [i\hbar^{-1}V_{ab}(z,t)\varrho_{ba} + \text{c.c.}] \tag{5.1.24}$$
$$\dot{\varrho}_{bb} = -\gamma_b(\varrho_{bb} - \varrho_{bb}^{(0)}) - [i\hbar^{-1}V_{ab}(z,t)\varrho_{ba} + \text{c.c.}]$$

where the dot represents $\partial/\partial t$, $V_{ab}(z,t) = \langle a \mid V(z,t) \mid b \rangle$ and $\varrho_{aa}^{(0)}$ and $\varrho_{bb}^{(0)}$ denote the population of levels a and b, respectively, in the absence of radiation. These equations play a role equivalent to that of last two eqns. (5.1.19). A difference is that here one can introduce a specific population relaxation rate for each atomic level, γ_a and γ_b (Fig. 5.1.1), instead of a unique rate $\gamma_\parallel$ for the population difference in eqns. (5.1.19). For the time evolution of the field in the case of a traveling plane wave, the first eq. (5.1.19) can be used.

5.2   Homogeneously Broadened Lasers

5.2.1   Theoretical Description

a. General Equations

The conceptually simplest kind of laser is that composed of an amplifying medium whose atoms (or molecules, ions, etc.) can be modeled as two-level systems with homogeneous broadening, and a ring cavity which supports unidirectional single-mode plane waves.

As described in the preceding section, this laser is described by eqns. (5.1.19), which represent a set of three complex or five real coupled equations of the type (3.1.2). When one of the parameters κ, $\gamma_\perp$ or $\gamma_\parallel$ is larger than the others, the differential equation for the corresponding variable can be adiabatically eliminated (Section 3.1.2.c). Thus, when $\gamma_\perp \approx \gamma_\parallel \gg \kappa$, only the first differential equation of eqns. (5.1.19) remains (class *A* lasers); when $\gamma_\perp \gg \kappa \approx \gamma_\parallel$ the first and third equations remain (class *B* lasers), and when $\gamma_\parallel \approx \gamma_\perp \approx \kappa$ the whole set of equations has to be solved (class *C* lasers). On the other hand, if the laser is tuned to resonance, then $E(t)$ and $P(t)$ can be chosen to be real (except for the exponential factor) and the set of eqns. (5.1.19) can be transformed into three real equations (as will be shown below).

As we are interested in dynamic behavior, we consider in this chapter the case of class *C* lasers, because it is the only one with three degrees of freedom (at least under resonance conditions) which are necessary for observing chaotic dynamics (Section 3.3.2). Pioneering work in the study of instabilities in the homogeneously broadened single mode laser were those of Grazyuk and Oraevskii (1964, already mentioned in Chapter 1), Haken (1966) and Risken et al. (1966).

First, let us transform eqns. (5.1.19) into a real form that makes a comparison with the Lorenz-model equations easier (Zeghlache and Mandel, 1985). To this end we define new dimensionless parameters:

$$\sigma = \frac{\kappa}{\gamma_\perp}, \quad b = \frac{\gamma_\parallel}{\gamma_\perp}, \quad r = \frac{N|g|^2 d_0}{\kappa \gamma_\perp} \tag{5.2.1}$$

where r represents the gain parameter (or pumping parameter or cooperation number), which is proportional to the atomic density N, to the pump strength d_0 and to $|\mu_{ab}|^2$ through $|g|^2$, and the new variables x_1, x_2, y_1, y_2 and z, defined by:

$$E(t) = \frac{\gamma_{\parallel}}{2ig}[x_1(\tau) + ix_2(\tau)]e^{-i\omega_L t}$$

$$gp(t) = \frac{g\,d_0}{2ir}[y_1(\tau) + iy_2(\tau)]e^{-i\omega_L t} \qquad (5.2.2)$$

$$d(t) = d_0\left[1 - \frac{z(\tau)}{r}\right]$$

where $\tau = \gamma_{\perp} t$ and ω_L represents the actual frequency of the field, which, because of pulling effects, is not exactly identical with the empty-resonator mode frequency ω_c (Sargent, Scully and Lamb, 1974; Haken, 1985a). In case of homogeneous broadening, ω_L is given by

$$\omega_L = \frac{\gamma_{\perp}\omega_c + \kappa\omega_{ab}}{\gamma_{\perp} + \kappa} \qquad (5.2.3)$$

and is independent of the field intensity. Equations (5.1.19) become

$$\dot{x}_1 = -\sigma(x_1 + \delta x_2 - y_1)$$

$$\dot{x}_2 = -\sigma(x_2 - \delta x_1 - y_2) \qquad (5.2.4)$$

$$\dot{y}_1 = -y_1 + rx_1 + \delta y_2 - x_1 z$$

$$\dot{y}_2 = -y_2 + rx_2 - \delta y_1 - x_2 z$$

$$\dot{z} = -bz + x_1 y_1 + x_1 y_2$$

where the dot now represents the derivative with respect to τ, and where the detuning δ between the field and resonator

$$\delta = \frac{\omega_L - \omega_c}{\kappa} = \frac{\omega_{ab} - \omega_c}{\gamma_{\perp} + \kappa} \qquad (5.2.5)$$

has been introduced.

b. Resonant Case. Lorenz-Haken Model

When the resonator is tuned to resonance ($\delta = 0$), one can choose solutions with $x_2 = y_2 = 0$ for all τ, so that eqns. (5.2.4) reduce to (by denoting $x_1 = x$, $y_1 = y$)

$$\dot{x} = -\sigma(x - y)$$

$$\dot{y} = -y + rx - xz \qquad (5.2.6)$$

$$\dot{z} = -bz + xy$$

which are exactly the Lorenz equations (2.1.1) (Haken, 1985b) that we considered in Section 2.1. There we showed the wide variety of unstable solutions that these simple equations yield for a certain range of values of the parameters σ, b and r. Other parameter ranges have been reported in, among others, Sparrow (1982) ($\sigma = 10$, $b = 8/3$) and Zeghlache and Mandel (1985) ($b = 1$, $\sigma = 3$). These results show qualitatively similar features to those in Section 2.1.

We recall that the condition for the stationary solution to become unstable under small perturbations is that the pump parameter r be larger than r_H ("second laser threshold"), which is given by eq. (2.1.4):

$$r > r_H = \sigma(\sigma + b + 3)/(\sigma - b - 1). \tag{5.2.7}$$

When r is continuously (adiabatically) increased and reaches the value r_H, a sub-critical Hopf bifurcation (Section 3.4.1) occurs, and the system falls into the basin of attraction of the chaotic Lorenz attractor (Section 2.1.3.a). The minimum possible value of r_H is 9 and its typical values lie around $10-20$. The existence of $r_H < \infty$ requires that the "bad-cavity" condition in eq. (2.1.5) or (2.1.6):

$$\sigma > b + 1 \quad \text{or} \quad \kappa > \gamma_\parallel + \gamma_\perp \tag{5.2.8}$$

be fulfilled for the laser system. The condition (5.2.8) is usually regarded as the necessary condition for the observation of unstable behavior. This notion may be misleading, however, because one has to distinguish between small and large perturbations of the stationary state (Casperson, 1985; Narducci et al., 1985). In the former case this state does not become unstable until r reaches the value r_H (we are assuming that r is being continuously (adiabatically) increased), whereas in the latter case the system may jump into the basin of attraction of the chaotic attractor for a value of r below r_H, ($r_A < r < r_H$; see Section 2.1.3) — this is sometimes referred to as "hard-mode" excitation; the larger the amplitude of the perturbations, the smaller is the value of r at which the jump may occur. Further, the existence of the chaotic attractor does not strictly require the bad-cavity condition (eq. (5.2.8)) to be fulfilled (Casperson, 1985).

The coexistence of the fixed-point attractor (eq. (2.1.3)) and of the chaotic attractor (Section 2.1.3) leads to hysteresis when r is first increased and then decreased: the change from stable to unstable emission when r is increased occurs for a larger value of r than the opposite change when r is decreased (generalized bistability, Section 3.3.3.c).

The chaotic behavior predicted by the Lorenz model, which is typically represented by Figs. 2.1.4 and 2.1.5, changes qualitatively when the ratio b/σ becomes small (Sparrow, 1982). Narducci et al. (1985) found that the chaotic behavior turns into a periodic self-pulsing regime for $\sigma = 4$ and $b \lesssim 0.2$.

An interesting question is why the laser emission becomes unstable when the pump parameter reaches the second laser threshold. The answer has been given by Sargent and co-workers (1983) (see Hendow and Sargent, 1982 and 1985, too) and Hillman et al. (1982), who showed that when a saturating single-mode field is excited it induces population pulsations (dynamic Stark effect) which produce net gain at side-bands separated by the Rabi frequency from the laser frequency and fulfil the cavity resonance condition just when the second threshold is reached. The simultaneous amplification of the laser and Rabi side-band fields gives rise to the unstable behavior. This phenomenon, in which several frequencies have a net gain and are simultaneously resonant with the cavity (i.e. they are associated with a single wavelength, which requires the presence of suitable dispersion effects), is known as "mode splitting". Several kinds of mode splittings which have different causes have been found in different types of lasers, which are described in the corresponding chapters.

Polarization effects which arise when one of the atomic levels is constituted by several magnetic sublevels were investigated by Puccioni et al. (1987). When two polarizations are active, a strong decrease in the second laser threshold with respect to the pure two-level case was found. The bad cavity condition is not required.

c. Non-Resonant Case

Zeghlache and Mandel (1983 and 1985) theoretically analysed the case $\delta \neq 0$, for the particular conditions $b = 1$, $\sigma = 3$, which fulfil the "bad-cavity" condition (7.2.8). As in the resonant case, the zero solution still exists, and the stationary solutions (2.1.3) now become

$$x_1^{\pm} = y_1^{\pm} = \pm \sqrt{b(r - 1 - \delta^2)}$$
$$x_2^{\pm} = 0, \quad y_2^{\pm} = -\delta x_1^{\pm} \tag{5.2.9}$$
$$z = r - 1 - \delta^2$$

which exist only for $r > 1 + \delta^2$ (first laser threshold). Both solutions correspond to the same stationary intensity $I = x_1^2 + x_2^2$. Keeping r as the control parameter, a linear stability analysis (Section 3.2.2.a) shows that the solutions (5.2.9) are stable for

$$r \leqslant r_{\mathrm{H}} = 3 + 16.8\,\delta^2 + (324 + 194.4\,\delta^2 + 262.44\,\delta^4)^{1/2}. \tag{5.2.10}$$

At $r = r_{\mathrm{H}}$, the system undergoes a Hopf bifurcation, which is subcritical for small detuning ($\delta^2 \ll 1/3$) and supercritical for large detuning ($\delta^2 \gg 1/3$). This is valid for $b = 1$ and arbitrary σ.

The results obtained by observing the time evolution in the $(x_1^2 + x_1^2, z)$ plane are summarized in Fig. 5.2.1. They were obtained by numerical integration of eqns. (5.2.4) with the initial conditions $x_1(0) = 0.01$, $y_1(0) = x_2(0) = y_2(0) = z(0) = 0$ and ignoring the initial transient period. For the resonant case ($\delta = 0$) the typical behavior already described in Section 2.1 is recovered: a sudden transition from stationary emission to chaos and the appearance of periodic windows (C^n represents an n-loop limit cycle) within the chaotic domain; for large r (not included in the figure), the system probably evolves out of chaos through a complete Feigenbaum scenario of period dedoublings (Section 3.4.2.b) similar to that described in Section 2.1.4 (this scenario is apparent in Fig. 5.2.1 for $\delta = 0.1$).

When the detuning δ increases, the unstable behavior becomes progressively simpler. The size of the chaotic domain diminishes and disappears at $\delta \simeq 0.25$. Only a periodic

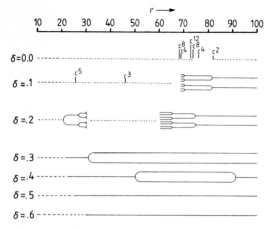

Fig. 5.2.1. Bifurcation diagram obtained by fixing the detuning and varying the pump parameter r. Dashed lines correspond to stable steady states, dotted lines to chaotic behavior and solid lines to periodic solutions: a single solid line for a one-loop limit cycle (C^1), a double line for a two-loop limit cycle (C^2), and so on (from Zeghlache and Mandel (1985)).

behavior of progressively simpler structure ($\ldots C^8 \rightarrow C^4 \rightarrow C^2 \rightarrow C^1$) remains. For $\delta \gtrsim 0.2$ the system no longer undergoes a sudden transition from a steady state to chaos (with increasing r), but changes to a periodic behavior. For $\delta \gtrsim 0.5$ only a simple periodic behavior C^1 appears, and for larger δ the steady states are stable for any value of r. Hence the introduction of a detuning destroys much of the complicated behavior predicted by the three equations of the Lorenz-Haken model and has a stabilizing effect. This suggests that instabilities appear, in general, when high gain and high losses are simultaneously present, so that a stationary equilibrium point between them is hard to reach.

d. Other Related Cases

When one of the control parameters in eqns. (5.2.6) is varied with time, the dynamical characteristics of the solutions can change. For instance, when the pump parameter r is increased linearly with time, the stability domain of the zero-intensity solution may be greatly increased (Mandel and Erneux, 1984). This corresponds to a dynamical stabilization of an unstable stationary solution.

An exact linear stability analysis of the plane-wave Maxwell-Bloch equations for a ring laser under conditions more general than the uniform-field limit was performed by Lugiato et al. (1986).

The case of standing-wave (i.e. with a Fabry-Pérot resonator) lasers has been mainly studied with respect to inhomogeneous broadening, which is treated in Section 5.3. For the case of homogeneous broadening a rate equation treatment was given for a new class of instabilities brought about by the interaction of a single resonator mode with a distributed feedback mode existing due to the susceptibility grating generated by the standing wave (Lawandy and Rabinovich, 1986).

The case of two coupled Lorenz lasers producing a common traveling wave field has been investigated (Lawandy, Plant and Lee, 1987; Lawandy and Lee, 1987). A lowering of the second laser threshold for both the chaotic and self-pulsing regimes and a coupling-induced transition from self-pulsing to chaotic behavior is predicted.

5.2.2   Experimental Results

As has been shown, the laser equations show instabilities in the case of

(a) $\quad \kappa > \gamma_{\parallel} + \gamma_{\perp}$ $\hspace{6cm}$ (5.2.11)

(b) $\quad r > r_{\mathrm{H}} = \dfrac{(\gamma_{\perp} - \gamma_{\parallel} + \kappa)(\gamma_{\perp} + \kappa)}{\gamma_{\perp}(\kappa - \gamma_{\perp} - \gamma_{\parallel})}.$

Equation (5.2.11a) has been termed the "bad-cavity" condition since it states that the field decay time of the empty laser resonator is shorter than the relaxation time of the active medium. In the frequency domain, this corresponds to a laser resonator linewidth larger than the homogeneous linewidth of the medium. In most popular lasers, such as solid-state lasers, CO_2 lasers or semiconductor-diode lasers, this condition is equivalent to a very high resonator loss which requires an extremely high pump parameter to reach even the first laser threshold. The pump parameter necessary to reach the second laser threshold given by eq. (5.2.11b), in the range of $10-20$ times the pump parameter at the first laser threshold, is then unattainable for these lasers. Hence there appeared to be little chance of experimentally observing the dynamics of the Lorenz equations. In fact, the Lorenz equations were dismissed in the discussion of laser instabilities as a "fascinating model without application to practical cases" (Casperson, 1983).

A closer inspection of eqn. (5.2.11), however, shows that the pump parameter necessary to reach the second laser threshold is proportional to the homogeneous linewidth of the medium. Thus, in lasers with narrow homogeneous linewidths one may expect to reach the second threshold with pump parameters technically within reach.

The homogeneous linewidth of a gas laser is determined by spontaneous emission and atomic/molecular collisions. The spontaneous emisson rate decreases with the optical frequency of the laser transition as v^3 (mode density of free space). Hence long-wavelength laser transitions effectively permit spontaneous emission to be eliminated. Evidently, the choice of special transitions with long lifetimes (metastable transitions) allows spontaneous emission to be eliminated also for short (visible) wavelengths. Collision rates are proportional to gas pressure. Therefore, long-wavelength infrared lasers operating at low pressure are one possible choice for observing laser dynamics. Estimates (Weiss et al., 1985) have shown that continously operating lasers in the far-infrared region which are pumped by a laser should permit the second laser threshold to be reached. An ammonia laser, e.g. at 100 µm wavelength, meets the "bad-cavity" condition with a resonator loss of only a few percent.

Critical voices have been raised, however, cautioning that these laser-pumped gas lasers are represented by nine instead of three equations and that it should be impossible to reduce these to the three Lorenz equations by any adiabatic elimination (Dupertuis et al., 1987; Ryan and Lawandy, 1987; Lawandy and Plant, 1986; Moloney et al., 1986; Uppal et al., 1987; Khandokin et al., 1988; Laguarta et al., 1988). See in this respect Sections 8.1 and 8.2. However, since the experimental results have been found to be in agreement in great detail, qualitatively and quantitatively, with the predictions of the Lorenz equations, there is little doubt that these lasers can represent homogeneously broadened single-mode lasers and show the corresponding dynamics. We therefore discuss the experimental results in this chapter.

Many theoretical models and also the Lorenz equations treat traveling waves, since the treatment of standing waves with their spatially varying intensity is exceedingly difficult. Therefore, the laser in the experiment has to be a ring laser. It has to emit in only one direction, a condition not usually easily fulfilled with ring lasers. The transitions in low-pressure far-infrared lasers are Doppler-broadened. However, since the excitation is by a laser, only one velocity group of molecules is excited. The active molecules are therefore all of the same resonance frequency and the active medium is then homogeneously broadened. Utilization of Doppler effects (technically by a proper choice and stabilization of the pump laser frequency) permits unidirectional emission of the ring laser.

The resonator used in the experiment is shown in Fig. 5.2.2. It is made up of three reflectors forming a triangular ring. One of them is a diffraction grating used in zeroth order as a plane far-infrared mirror and simultaneously in first order to couple the pump laser beam into the resonator volume. A gold-mesh reflector of 30 µm grid constant is used as a semitransparent mirror to couple out the generated 81 µm wavelength radiation. The resonator is unusual for far-infrared lasers. It is designed to generate unperturbed TEM_{00} modes, which are the best possible approximation to a plane wave described by the Lorenz equations. The spatial filter-diaphragm combination permits the pump laser beam power to be changed without changing its geometry, direction or frequency, which is necessary to vary the laser pump parameter in a controlled manner.

With a comparatively elaborate two-laser set-up, it is possible, as necessary, to set and maintain the pump laser frequency within 50 kHz. Details are described in Chapter 8.

In agreement with estimates (Weiss et al., 1985), the 81 µm ammonia laser shows self-pulsing for ratios of resonator linewidth to homogeneous linewidth larger than ca. 2.5. If the ammonia pressure is increased, increasing the homogeneous linewidth and thus decreasing the linewidth ratio below 2.5, no pulsing occurs, irrespective of a further increase in pump power. The pulsing, however, reappears if the resonator loss is increased so that the linewidth ratio is again larger than 2.5 even though in this case the laser first threshold is higher and with it the second threshold. This is a clear indication that the instability occurs only under "bad-cavity" conditions.

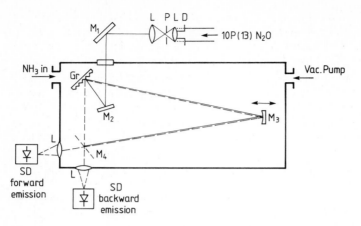

Fig. 5.2.2. Ammonia ring laser used for the experiments. The ring consists of three reflectors: grating, curved gold mirror and semi-transparent (gold-mesh) mirror. The first order of the grating serves to couple the pump laser beam into the resonator, the zero-order reflection of it is utilized for the generated radiation whose wavelength is long compared with the grating period. A combination of spatial filter and diaphragm serves as a pump intensity attenuator to vary the pump strength. M: mirror, L: lens, P: pin-hole, Gr: grating, SD: detector diode, D: diaphragm.

Fig. 5.2.3. The chaotic motion of the Lorenz attractor. The two centers of the spirals correspond to continuous laser emission. The attractor is a quasi-two-dimensional object in the E-, P- and D-space.

An overall view of the dynamics of the Lorenz equations is shown in Fig. 5.2.3 in the three-dimensional parameter space (inversion, polarization and field) of the Lorenz equations (see also the detailed example in Chapter 2). The system's motion consists of spirals around two saddle focus points and jumps between the two basins of attraction of the focus points. The focus points are stable below the second laser threshold and correspond to the continuous laser emission, for which the field can have two different phase values differing by 180°, due to the symmetry of the laser equations.

If the field is observed, the resulting time dependence is as shown in Fig. 5.2.4. It reflects the spiral motion around the two saddle points. The length of the spirals varies randomly. The corresponding intensity-time dependence is obviously the square of the field (Fig. 5.2.5).

The growing spirals and their irregular length are characteristic of the Lorenz model. For comparison, Fig. 5.2.6 shows the intensity of the 81 μm ammonia laser measured when the spectrum of the intensity is continuous (Fig. 5.2.15) and thus the laser emission may be considered to be chaotic.

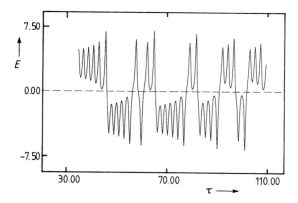

Fig. 5.2.4. Laser field strength as function of time calculated from the Lorenz equations.

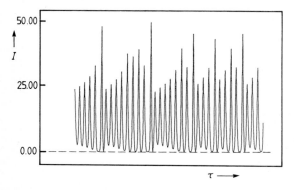

Fig. 5.2.5. Laser intensity as function of time corresponding to Fig. 5.2.4.

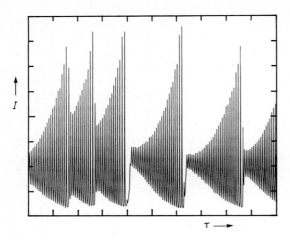

Fig. 5.2.6. Pulsing of the ammonia laser measured for central resonator tuning. The irregular lengths of the "pulsing spirals" are clearly recognizable.

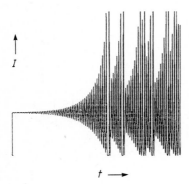

Fig. 5.2.7. Destabilization of the steady-state laser emission calculated for the Lorenz model.

Fig. 5.2.8. Destabilization of steady-state laser emission (measured).

Figure 5.2.7 shows the calculated intensity after the pump is suddenly switched to a value above the second threshold. The continuous steady-state emission becomes unstable, which results in the system point spiralling away from the steady-state point. Reaching the boundary at which the system point jumps to the basin of attraction of the other fixed point then starts the chaotic time evolution. For comparison, Fig. 5.2.8 shows the intensity of the 81 µm ammonia laser after the pump is increased from below the second threshold to 5% above the second threshold. The destabilization of the steady-state emission is clearly as in the calculation (Fig. 5.2.7).

A characteristic feature of the Lorenz chaos is also the comparatively high second threshold. In inhomogeneously broadened lasers (Casperson, 1978), for example, the second threshold is typically within twice the first threshold; in fact, it may be as low

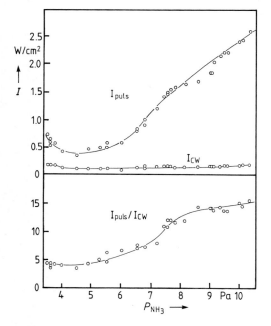

Fig. 5.2.9. Instability threshold (onset of pulsing) of the ammonia laser as a function of ammonia pressure. Top: threshold pump intensity for continuous laser emission (I_{cw}) and pulsing laser emission (I_{puls}). Note that $\gamma_\perp$ is proportional to pressure in this laser. Bottom: ratio of second to first laser threshold, as function of ammonia pressure. The ratio of second to first threshold calculated from the Lorenz equations is ca. 14.

as the first threshold itself. In contrast to these cases, the second threshold of the Lorenz model is typically $10-20$ times the first. From published relaxation data on ammonia (Lawandy and Plant, 1986), one calculates a second threshold at $13-14$ times the first threshold for the 81 µm ammonia laser.

Figure 5.2.9 shows the second threshold (onset of pulsing) as measured on this laser as a function of gas pressure. In the low-pressure range the second threshold is found to be substantially lower than $13-14$ times the first. This shows that the dynamics is not that of the Lorenz model. In this range, the chaotic pulse train of the laser output is also not of the "spiral" form predicted by the Lorenz equations. One would in fact expect three-level coherences, which make the system different from the Lorenz model, to be strongest at low pressure.

In the high-pressure range (ca. 9 Pa), however, we find the second threshold value to be as the Lorenz equations predict and it is in this pressure range that the chaotic pulsing is of the "growing spiral" type typical of the Lorenz model.

To appreciate better the similarity between measured chaotic pulses and the predictions of the Lorenz model, Fig. 5.2.10b shows a phase space projection of the laser emission constructed from measured laser pulses. The laser field strength E is plotted against the first time derivative of the field strength, dE/dt. For comparison, the same phase space projection has been calculated for the Lorenz equations (Fig. 5.2.10a). The similarity is obvious.

Since the intensity is the square of the field strength, the question of whether the attractor underlying the motion of the system has two sheets (corresponding to the unstable manifolds of the two saddle fixed points) or only one cannot be decided from intensity measurements alone. In the former instance the field changes sign after each spiral, but in the latter instance it does not.

To answer this question, measurements of the field rather than the intensity are necessary. These were carried out by a heterodyne technique (Weiss et al., 1988). Figure 5.2.11 shows the measured spiral-intensity pulsing together with the in-phase and the $90°$ out-of-phase component of the optical field. Measurement of both of these components is necessary to detect possible phase changes much smaller than π, but also for the technical reason that the heterodyne reference frequency cannot be set to any "natural" laser frequency such as the gain line center (since the latter is not known). Consequently, a phase drift has to be expected simply owing to possible detuning of the reference frequency and the laser. The two components of the electric laser field

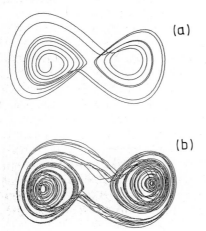

(a)

(b)

Fig. 5.2.10. Phase-space projection in the $E - \dot{E}$ plane. (a) Calculation of Lorenz model; (b) measurement from ammonia laser.

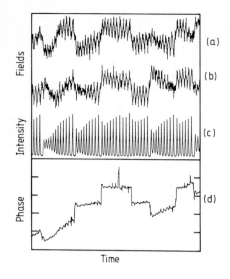

Fig. 5.2.11. Symmetric Lorenz attractor. (a) In-phase component, (b) out-of phase component, (c) intensity, and (d) phase of the electric field of a laser (unit of phase is π). Measurement shows sign reversal of the field at the end of "spirals". Changes of the field sign are seen as changes of phase by π.

thus measured permit in principle the reconstruction of both field magnitude and phase as a continuous function of time.

It is clearly recognizable from Fig. 5.2.11(a)−(c) that the sign of the electric field reverses at the beginning of each new spiral, as would be expected for the Lorenz attractor with its two fixed points of opposite sign.

Figure 5.2.11(d) shows the change of 180° in the field phase at the beginning of each spiral more clearly. The phase has been reconstructed from the in-phase and 90° out-of-phase field components measured. The plot clearly shows the phase jumps of magnitude 180°. The superposed slow change of phase is due to detuning of the reference frequency from the laser frequency and to laser frequency and reference frequency "jitter" during the measurement time.

Obviously, the motion on the Lorenz attractor can become periodic. This happens for all parameter values of the pump, e.g. when the time between two jumps between attractor "leaves" is an integer multiple of the time required to move around the fixed point once. Figure 5.2.12 shows such a periodic case. After three revolutions around one fixed point, the system jumps to the other fixed point (the other leaf of the attractor) where it executes another three revolutions, then jumps back, and so on.

Figure 5.2.13 shows the case where six revolutions occur between jumps. Figures 5.2.12 and 5.2.13 clearly indicate the two-sided geometry of the attractor, as do Figs. 5.2.11a−d, in agreement with the Lorenz attractor.

So far, all experimental tests have addressed predictions of the genuine (or real) Lorenz equations. In a real laser, the parameter most easily controllable is the laser resonator tuning. The Lorenz equations extended for detuning (the complex Lorenz equations; see Zeghlache and Mandel, 1985) also predict characteristic features which can be tested experimentally. Among them are a period-doubling transition to chaos when the pump is set above the second laser threshold and the laser resonator is tuned towards the gain line center (Fig. 5.2.1).

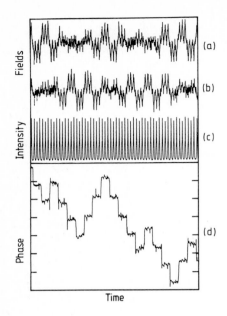

Fig. 5.2.12. Same as Fig. 5.2.11, except the dynamics is periodic.

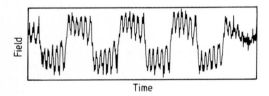

Fig. 5.2.13. Electric field component showing a symmetric period-6 attractor.

Relatively close to the line center, a weakly chaotic "period-three" attractor is predicted (weakly chaotic meaning that the width over which the almost periodic trajectories are spread is small in comparison with diameter of the trajectories; see Fig. 5.2.14), which is then followed, when tuning completely to the line center, by the Lorenz attractor (Zeghlache and Mandel, 1985).

The experimental results are shown in Figs. 5.2.15 and 5.2.16. When tuned far enough away, the laser operates continuously. Closer to the line center self-pulsing starts, followed on further tuning towards the line center by the period-doubling cascade (Fig. 5.2.15 left hand column) recognizable by the subharmonics in the intensity spectrum. As predicted by the logistic map, the period-doubling cascade is followed by the part-chaotic inverse cascade (Fig. 5.2.15 right hand column) until full chaos is reached.

Further tuning results in "period-three" pulsing (Fig. 5.2.16) where the varying amplitudes of the pulses can be interpreted as the weak chaos expected from the complex Lorenz equations. Figure 5.2.17 shows a mixture of Lorenz spirals with "pe-

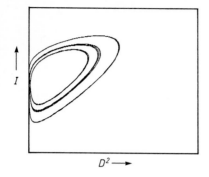

Fig. 5.2.14. Weakly chaotic P-3 attractor calculated from the Lorenz equations with detuning $\delta = 0.14$, represented in the intensity-inversion plane.

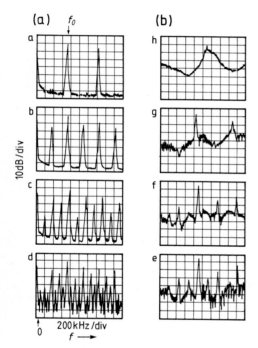

Fig. 5.2.15. Intensity spectra of ammonia laser measured when tuning progressively (a to h) the laser resonator towards the gain line center. The left column shows the period doubling cascade and the right column the "reverse" noisy cascade.

riod-three" pulsing which indicates that the weakly chaotic "period-three" attractor and the Lorenz attractor are in fact neighboring in parameter space. Tuning completely to the line center results in the spiral chaos of the Lorenz attractor.

As far as tests of overall properties are concerned, intensity-time series, autocorrelation functions and phase plots can be compared with the corresponding properties calculated from the Lorenz equations. Figure 5.2.18b shows a time series of pulses calculated from the Lorenz equations for parameters realistic for the ammonia laser. Figure 5.2.18a shows a measured time series of pulses from the ammonia laser. Note

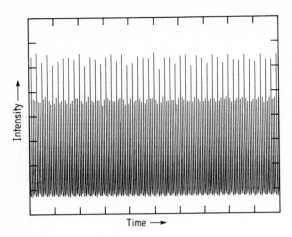

Fig. 5.2.16. Weakly chaotic period-3 pulsing of slightly detuned ammonia laser. Chaoticity shows up in the pulse height corresponding to Fig. 5.2.14.

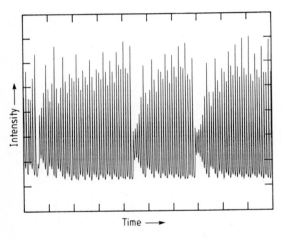

Fig. 5.2.17. Measured mixture of *P*-3 pulsing and Lorenz-type "spiral" chaos, showing that the Lorenz attractor and the *P*-3 attractor are neighboring in parameter space.

the similarity between the calculated and measured pulses even of details such as the last large pulse terminating each spiral.

The autocorrelation function of the calculated pulsing intensity reveals that the chaotic dynamics of the Lorenz equations are still largely periodic. This is reflected by the periodicity of the autocorrelation function (Fig. 5.2.19). Notable are the "revivals" at long delay times. The autocorrelation function of the Lorenz attractor has no monotonously decreasing envelope, such as a stochastic function would have. The increase in the autocorrelation function for long delay times (the "revivals") reflects the deter-

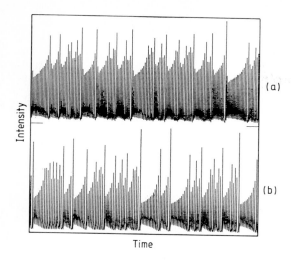

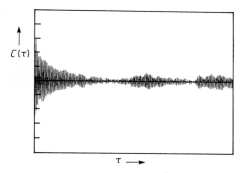

Fig. 5.2.18. (a) Measured intensity pulses of the ammonia laser to be compared with (b). (b) Intensity pulses calculated from the Lorenz equations with $r = 15$, $b = 0.25$, $\sigma = 2$.

Fig. 5.2.19. Intensity autocorrelation function for the Lorenz model.

ministic nature of chaos. Once a trajectory comes close to a point it has visited before, it will for some time (the Lyapunov time) develop similarly to the time before ("recurrence").

Note the similarity of the autocorrelation function calculated from the Lorenz equations (Fig. 5.2.19) and that calculated from the measured pulses of the ammonia laser (Fig. 5.2.20).

Another way of looking at attractors is by their phase-space representations. These can be obtained, e.g. by plotting the nth point of the time series against the $(n + \tau)$th point. Depending on the choice of τ, the plot obtained will look different. Choosing τ to be very small, the nth and $(n + \tau)$th points will be strongly correlated. In this case the points of the plot will lie on or close to a 45° straight line. If τ is chosen to be very large, the points will have no correlation at all and a structureless cloud of points results. For intermediate values of τ (typically of the order of one tenth to one quasi-period), the plot shows "structure" reflecting the structure of the n-dimensional

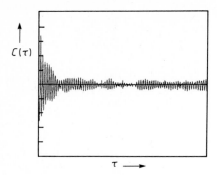

Fig. 5.2.20. Intensity autocorrelation function for the ammonia laser.

(a) LORENZ MODEL   (b) LASER

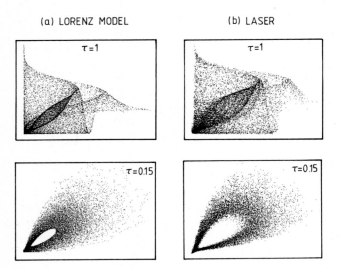

Fig. 5.2.21. (a) Phase-space projections of Lorenz model intensities for $\tau = 0.15$ and 1 pulsing period; (b) same for measured intensity of ammonia laser.

attractor underlying the system dynamics. Figure 5.2.21a shows phase-space plots for two different values of τ calculated from the time series (Fig. 5.2.18b) of the Lorenz equations. For comparison, the phase-space plots calculated for the same τ values from the measured intensity of the ammonia laser (Fig. 5.2.18a) are given in Fig. 5.2.21b (Weiss et al., to be published). Note again the similarity between the laser and Lorenz equation results, even in detail.

It should be noted that the use of this sort of phase-space representation allows one to recover some of the structure of the attractor only if the latter is of relatively low dimension. The Lorenz attractor is embedded in a three-dimensional space and hardly of dimension larger than two. In experiments involving four-dimensional attractors, it is found that structure is hardly ever evident in phase-space plots, irrespective of delay time (Klische et al., in print).

From such delayed time series, finally the fractal dimension of the attractor can be calculated, as is described in Chapter 4. For details see Section 4.3, where we have noted the excellent agreement in the dimensions of calculated intensities from the Lorenz model and the values measured with the ammonia laser.

The ammonia laser represents an example of a large class of lasers, viz. low-pressure gas lasers optically pumped by lasers. The number of laser transitions discovered so far in this class is around 3000, and one can estimate that ca. 10^6 actually exist. Even though not all of them may have transition dipole moments large enough to reach 20 times the first laser threshold without introducing 3-level coherences, it appears that the simple dynamics of the Lorenz model should be characteristic of this large class of lasers. Initial experiments on mid-infrared and visible lasers (vibrational and electronic transitions) have indeed shown chaotic dynamics with characteristics similar to those of the ammonia laser described (Siemsen et al., 1987; Wu and Weiss, 1987).

5.3 Inhomogeneously Broadened Lasers

Inhomogeneously broadened lasers have been extensively studied both theoretically and experimentally and qualitatively and sometimes quantitatively agreeing results have been achieved.

5.3.1 Theoretical Description

Inhomogeneous broadening in gas lasers is brought about by the atomic velocities and the resulting Doppler effect. Let us consider, as before, a uni-directional single-mode gas laser. In the form (5.1.19) of the laser equations, the polarization $p(t)$ and population inversion $d(t)$ now become velocity dependent, so that one has to write a pair of equations like the last two eqns. (5.1.19) for each atomic velocity class and to sum up in the last term of the first equation all the contributions to the total polarization coming from the different velocity classes (Haken, 1966):

$$\dot{E}(t) = -(i\omega_c + \kappa)E(t) + Ng \int_{-\infty}^{\infty} p(v,t)W(v)dv$$

$$\dot{p}(v,t) = -[i\omega_{ab}(v) + \gamma_\perp]p(v,t) + gE(t)d(v,t) \tag{5.3.1}$$

$$\dot{d}(v,t) = -\gamma_\parallel [d(v,t) - d_0(v)] - 2g[E(t)p(v,t)^* + \text{c.c.}]$$

where v is the atomic velocity along the resonator axis z and $W(v)$ is the Maxwellian distribution:

$$W(v) = \frac{1}{\sqrt{\pi}\,u}\,e^{-\frac{v^2}{u^2}} \tag{5.3.2}$$

where $u = (2k_B T/m)^{1/2}$ represents the most probable atomic velocity and the Doppler width is given by $\sigma_D = ku/\sqrt{2}$. In the second equation, $\omega_{ab}(v)$ represents the transition frequency (in the laboratory frame) for an atom moving with velocity v:

$$\omega_{ab}(v) = \omega_{ab} + kv \qquad (5.3.3)$$

where k represents the wavenumber. The population difference in the absence of fields, $d_0(v)$, is now v-dependent. A v-averaged population difference can be defined as

$$d_0 = \int\limits_{-\infty}^{\infty} d_0(v) \cdot W(v)\mathrm{d}v . \qquad (5.3.4)$$

The laser equations can alternatively be expressed in terms of the density-matrix elements, as in eq. (5.1.21) or (5.1.24). Owing to the atomic motion, the ensemble-averaged density matrix now becomes v-dependent: $\varrho(z,v,t)$. To deal with this latter dependence, a spatial derivative $v\partial/\partial z$ has to be added to $\partial/\partial t$ in eqns. (5.1.21) and (5.1.24):

$$\frac{\partial}{\partial t} \longrightarrow \frac{\partial}{\partial t} + v \cdot \frac{\partial}{\partial z} \qquad (5.3.5)$$

and the polarization (5.1.23) is now

$$\mathscr{P}(z,t) = \frac{N}{V}\mu_{ab} \int\limits_{-\infty}^{\infty} \varrho_{ab}(z,v,t)\,W(v)\mathrm{d}v . \qquad (5.3.6)$$

Let us consider, for instance, the form (5.3.1) of the laser equations. In contrast to the previous case of homogeneous broadening, they now represent a large set of coupled equations. For instance, if one takes n_v atomic velocity classes for performing numerically the sum in the first of eqns. (5.3.1), the total number of coupled complex equations is $2n_v + 1$. In practical cases for n_v one has taken values of the order of $10-100$ (Casperson, 1978). As described below, in spite of this large number of degrees of freedom, low-dimensional chaos occurs in inhomogeneously broadened lasers.

Casperson (1978) demonstrated both experimentally and theoretically the appearance of pulsed emission under cw pumping in a xenon laser, obtaining good qualitative agreement between both kinds of results. The numerous theoretical analyses performed since then by different workers can be divided into two groups: those which analyse the steady-state solutions and their stability against small perturbations, and those which obtain (numerical) solutions corresponding to conditions above the laser threshold.

Within the first group, Casperson (1980 and 1981) and Sargent and co-workers (1983) gave stability analyses of the laser equations based on the weak-sideband approach (Section 3.2.2.b). Sargent and co-workers (1983) additionally treated the strong signal case, whereas Mandel (1983 and 1985), Lugiato et al. (1985) and Abraham et al.

(1985) studied the steady-state solutions by means of the standard linear stability analysis (Section 3.2.2.a). Abraham et al. (1985) gave a wide range of results, including analytical expressions for the steady-state solutions and their stability analysis for both line-center and detuned cases. Figure 5.3.1 shows typical curves for the steady-state intensity and laser frequency as a function of the resonator detuning (they correspond to a Lorentz-broadened case instead of a Gaussian one, but the qualitative behavior is similar in both instances). In general, a unique solution with finite intensity exists, but for large gain and detuning (wings of curves 1 and 2 in Fig. 5.3.1.a) two steady-state solutions (and in other cases three steady-state solutions) appear. Bistability between zero-intensity and finite intensity solutions is possible and appears to be experimentally accessible. The steady-state solutions are independent of the value of $\gamma_\parallel/\gamma_\perp$, but their stability character is influenced by this value. Figure 5.3.1 shows the strong mode-pulling effect on the laser frequency ω_L. Unlike the homogeneously broadened laser, here this effect is intensity dependent.

The phase diagram in Fig. 5.3.2 shows the different regions of stable or unstable steady-state solutions in the detuning-versus-gain plane, obtained through a linear stability analysis. The system undergoes transitions from stable to unstable steady states through Hopf bifurcations (Section 3.4.1), as in the case of homogeneous broadening. Here however two characteristic exponents can assume a positive real part, giving rise to the domain frontiers defined by the curves 1 and 2. Figure 5.3.2 corresponds to a particular case, but nevertheless it shows some of the typical features of the inhomogeneously broadened lasers. The most important one is that the instability threshold is very close to the laser threshold, in contrast to homogeneously broadened

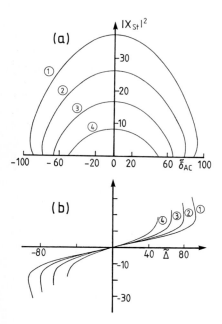

Fig. 5.3.1. (a) Steady-state output intensity $|x_{st}|^2$ (normalized to the saturation intensity) for a Lorentz-broadened laser as a function of the cavity detuning $\tilde{\delta}_{AC} = (\omega_{ab} - \omega_c)/\gamma_\perp$, for $\sigma_D/\gamma_\perp = 10$, $\kappa/\gamma_\perp = 20$ and for a decreasing gain parameter from curves 1 to 4. (b) Corresponding values of the steady-state offset frequency $\tilde{\Delta} = (\omega_{ab} - \omega_L)/\gamma_\perp$. (From Abraham et al. (1985)).

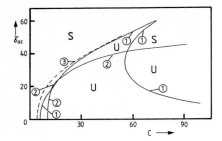

Fig. 5.3.2. Phase diagram showing the stable (S) and unstable (U) regions for the steady-state solution in the detuning versus gain plane (C, $\tilde{\delta}_{AC}$). C is the gain or pumping parameter ($C \propto N d_0 |g|^2$), $\tilde{\delta}_{AC} = (\omega_{ab} - \omega_c)/\gamma_\perp$, $\sigma_D/\gamma_\perp = 10$, $\kappa/\gamma_\perp = 5$ and $\gamma_\parallel/\gamma_\perp = 2$. Curves 1 and 2 limit the different domains and curve 3 represents the position of the first laser threshold. (From Abraham et al. (1985)).

lasers where the gain necessary for such thresholds differs by a factor of 9 or more. If the Doppler width σ_D is increased, both thresholds move progressively closer to each other (both of them diverging as σ_D diverges). This means that instabilities may appear just after the appearance of laser oscillation.

Another important result of the stability analysis in the case of Gaussian broadening is that a "bad-cavity" condition $\kappa > \gamma_\perp$ is necessary for the appearance of instabilities, as with homogeneously broadened lasers.

When the gain parameter is increased beyond the instability threshold, the weak-sideband approach or the linear stability analysis are no longer meaningful and in general exact time-dependent solutions have to be obtained by numerical methods. This has been carried out by different workers.

Bandy et al. (1985) studied the time-dependent solutions that arise when the pump parameter is progressively increased beyond the laser threshold. The same was done by Tarroja et al. (1986) for a set of parameters in an attempt to match the experimental conditions of a uni-directional single-mode He-Xe ring laser. Figure 5.3.3 shows the different domains of unstable behavior that have been found for resonant tuning. Just above the instability threshold, the laser intensity shows weak oscillations about its mean value that correspond to the limit cycle which has been created as a result of the supercritical Hopf bifurcation. With increasing gain the limit cycle increases in size and becomes the asymmetric attractor of Fig. 5.3.4.a, which characterizes zone A in Fig. 5.3.3.b; the laser intensity shows a regular train of pulses. On increasing the gain further, the emission becomes progressively chaotic (zone C' in Fig. 5.3.3.b). Within this region, windows with symmetric periodic solutions appear, indicating that a gradual transition towards a symmetric attractor is taking place. Indeed, after the chaotic domain a region of weakly asymmetric solutions corresponding to the attractor in Fig. 5.3.4.b appears; the asymmetry of the solutions progressively decreases and eventually the attractor becomes fully symmetric (zone S in Fig. 5.3.3.b). If the gain is continuously reduced (adiabatic return), a reverse "symmetry-breaking" transition (i.e. from sym-

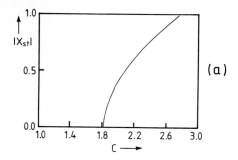

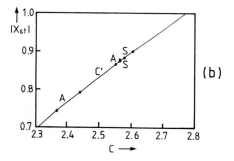

Fig. 5.3.3. (a) Modulus of the steady-state field $|x_{st}|$ for a Gaussian-broadened laser in resonance as a function of the gain parameter $C \propto N |g|^2$, for $\sigma_D/\gamma_\perp = 3.72$, $\kappa/\gamma_\perp = 4.67$ and $\gamma_\parallel/\gamma_\perp = 0.183$. (b) Expanded region of the curve in (a), showing the different domains of unstable behavior: A, asymmetric attractor; C', chaotic attractor; S, symmetric attractor; A & S, region of coexistence of both asymmetric and symmetric attractors. (From Tarroja et al. (1986)).

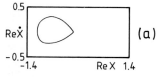

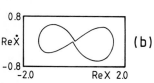

Fig. 5.3.4. Projection of the limit-cycle attractor in the $(\text{Re}X, \text{Re}\dot{X})$ plane (X represents the field amplitude) for two regions of unstable emission in Fig. 5.3.3: (a) $C = 2.3677...$ ($|x_{st}| = 0.79$), asymmetric attractor; (b) $C = 2.56622...$ ($|x_{st}| = 0.875$), weakly asymmetric attractor. (From Tarroja et al. (1986)).

metric to asymmetric solution) takes place and a region of overlap between symmetric and asymmetric solutions is found (zones A and S in Fig. 5.3.3.b).

Figure 5.3.5 shows the heterodyne power spectra of the laser output corresponding to the different solutions described above, found when the gain is increased as in Fig. 5.3.3.b. Heterodyne power spectra can be obtained experimentally in a way that is described below, and provide the spectrum of the laser field (instead of the spectrum of the laser intensity provided by the usual intensity detection). The central peak in Fig. 5.3.5.a marks asymmetric solutions (such as Fig. 5.3.4.a), whereas the spectrum in Fig. 5.3.5.f, with no central peak, marks symmetric solutions. The spectra in Figs. 5.3.5.b — d show a broad-band component corresponding to the chaotic behavior.

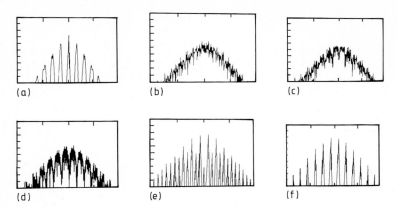

Fig. 5.3.5. Numerically obtained heterodyne power spectra of the laser emission for the same conditions as in Fig. 5.3.3. Spectra are shown as the excitation C increases from (a) to (f). A unit corresponds to 37.6 MHz (horizontal axis) or 6.5 dB (vertical axis). (From Tarroja et al. (1986)).

When the laser is off-resonance, the results are similar, with a greater tendency towards asymmetric attractors. The chaotic domain diminishes with increasing detuning, suggesting a tendency for more regular behavior, as in the case of homogeneously broadened lasers (Section 5.1).

By selecting suitable (and realistic) parameter values, Shih and co-workers (1985) and Ackerhalt and co-workers (1985) found the three most popular routes to chaos when a parameter is varied (Section 3.4.2): period-doubling (Feigenbaum), two-frequency (Ruelle-Takens) and intermittency (Pomeau-Manneville). Graham and Cho (1983) developed approximate methods which are commented in Mandel (1983 and 1985) and Bandy et al. (1985).

It is also worth mentioning that the time-averaged laser intensity is larger in the actual pulsing regime than in the (unstable) steady-state regime corresponding to the same set of parameter values (Casperson, 1985).

No instabilities are found with a rate equation model that ignores coherent and dispersion effects. We finally report briefly the semi-intuitive physical interpretation for the pulsing instability given by Casperson (1980 and 1981) in terms of an "induced mode splitting" (Maeda and Abraham, 1982).

The most prominent effect of a coherent field on an inhomogeneously broadened ensemble of atoms, be they absorbing or inverted, is "hole burning" (Fig. 5.3.6). Within a range of atomic frequencies as wide as the homogeneous linewidth, the atomic population is substantially changed, resulting in reduced absorption or amplification. This change in population distribution brings about a variation in the refractive index of the medium (Fig. 5.3.6). The phase condition for laser oscillation requires that the optical length of a resonator round trip be an integer number of wavelengths. With the refractive index profile (Fig. 5.3.6) this condition is fulfilled for several frequencies. Since there is gain at these additional frequencies, laser emission is possible ("mode splitting"). The laser can therefore emit at several frequencies which are phase-locked

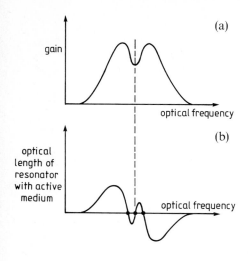

(a)

gain

optical frequency

(b)

optical
length of
resonator
with active
medium

optical frequency

Fig. 5.3.6. (a) Lamb dip and (b) saturation dispersion in an inhomogeneously broadened gain line. For the three marked points in (b) the wavelengths are the same although the frequency differs. Simultaneous oscillation of the three frequencies is possible since they all have gain.

through their close interaction. The result is mode-locked pulsing. This picture applies only to inhomogeneously broadened lasers and the pulsing of homogeneously broadened lasers has to be pictured in a different way (Section 5.1).

When a laser works with a Fabry-Pérot cavity instead of a uni-directional ring cavity, several new features appear because a standing wave is generated instead of a travelling wave. Casperson (1978, 1985 and 1988) performed a detailed theoretical analysis in which standing-wave effects, internal relaxation and velocity-changing collision effects were taken into account. As is shown in the next section, the results are in very good agreement with the experimental observations. Other analyses of the standing-wave case have been reported (Hendow and Sargent, 1985; Englund, 1986 and references cited therein; Gioggia et al., 1988).

In Section 8.2.2, a model of a laser with possibly the simplest kind of inhomogeneous broadening is considered. The laser medium has only two resonant atomic frequencies, instead of a continuous distribution of atomic frequencies (brought about by the Doppler effect) as in the present case.

5.3.2 Experiments

The only types of lasers broadened inhomogeneously by the Doppler effect and showing enough gain to fulfil the requirement of reaching a high enough pump parameter under "bad-cavity" conditions to reach an instability threshold appear to be the He-Xe laser at 3.5 μm, on which consequently most experiments have been done, and the He-Ne laser at 3.39 μm. These two systems show outstandingly high gain, larger for the He-Xe than the He-Ne system.

a. He-Xe laser

This laser was the first continuous laser to be studied thoroughly in view of pulsing instabilities. Casperson found experimentally that He-Xe lasers, instead of providing continuous output, showed pulsing phenomena (Casperson, 1978). A typical experimental pulse train is shown in Fig. 5.3.7. The pulses appear as a principal leading pulse followed by "ringing" echo pulses. The theoretical model developed by Casperson exactly reproduces this kind of result (Casperson, 1985 and 1988) (Fig. 5.3.8).

For central laser tuning, when the pump strength is decreased different pulsation forms are encountered (Fig. 5.3.3). A plot of the pulsing frequency as a function of pump parameter is shown in Fig. 5.3.9, with the experimental pulsation frequencies indicated. On the grounds of repetitivity of the pulses, conclusions were drawn as to whether the pulses are periodic, period-doubled or chaotic. In general, the agreement between experimental and theoretical pulse shapes and frequencies is very satisfactory, in spite of the fact that the experiments were done on standing-wave lasers whereas the theoretical studies considered traveling-wave beams.

Chaotic dynamics is concluded to occur always in the regions where the pulsation forms change qualitatively. The picture is suggested that two pulsation forms in the range between their "pure" existence compete, thus producing chaos.

A large body of experimental data on instabilities of He-Xe lasers has been provided by Hoffer et al. (1985).

One of the questions addressed is the experimental proof of chaotic dynamics. Today still the most convincing proof is a demonstration of one of the well known "transitions to chaos" as discussed in Chapter 3, such as the period doubling. Gioggia and Abraham (1983a) were able to find conditions in which a broad-band noise spectrum of the

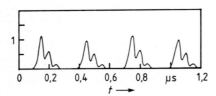

Fig. 5.3.7. Experimentally measured pulses of a Xenon laser with continuous excitation by a 50 mA discharge current (Casperson, 1978, 1980 and 1981).

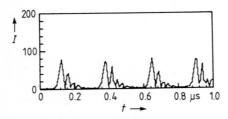

Fig. 5.3.8. Pulses as calculated for a Xenon laser model (Casperson, 1978, 1980 and 1981). Note the good agreement with experiment (Fig. 5.3.7).

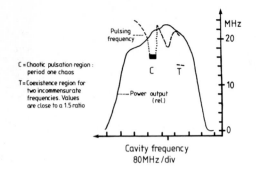

C = Chaotic pulsation region :
period one chaos

T = Coexistence region for
two incommensurate
frequencies. Values
are close to a 1.5 ratio

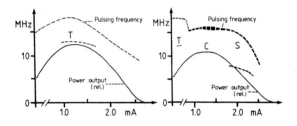

Fig. 5.3.9. Power output and laser pulsing frequency as a function of discharge
current and resonator tuning for the
3.5 μm He-Xe laser (Abraham et al.,
1983).

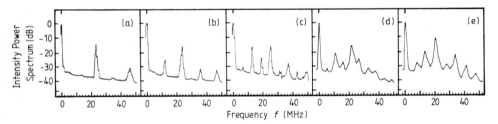

Fig. 5.3.10. Spectra of He-Xe laser intensity for different laser resonator detunings. A clear period-
doubling cascade appears with period-4 chaos (d) and period-3 chaos (e) (Gioggia and Abraham,
1983a).

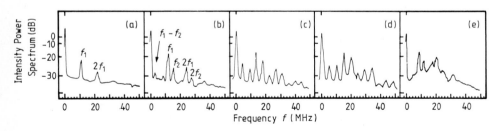

Fig. 5.3.11. Spectra of He-Xe laser intensity for fixed resonator tuning and increasing discharge
current (gain). The spectrum (b) consists of all linear combinations of f_1 and a second frequency
f_2, which lock at a frequency ratio of 3/4 (c). Finally, the space between the lines is filled by noise,
indicating chaotic emission (Gioggia and Abraham, 1983a).

output intensity of a standing-wave laser is preceded by two period doublings (Fig. 5.3.10), by the onset of two periodic pulsing frequencies (Fig. 5.3.11) and by a successive broadening of initially sharp lines in the intensity spectra (Fig. 5.3.12). The broadening was interpreted as an increasing number of random bursts typical of the intermittent onset of chaos. No relation with theoretical models was found. The experiment simply shows that these three popular routes to chaos do occur with this laser, similarly to the theoretical work of Shih et al. (1985).

Experimental conditions that can be more easily related to theoretical treatments are realized with a ring laser supporting only one traveling wave mode (Hoffer et al., 1985). As the pump is increased for line-center tuning, different pulsing frequencies are observed (Fig. 5.3.13). The continuous operation near the laser threshold is followed by periodic pulsing. Above this, lines of half of the pulsing frequency are observed. This is followed by a region characterized by two pulsing frequencies. When the laser resonator detuning from the gain line center is varied, periodic and chaotic pulsing is found (Fig. 5.3.14), not entirely different from what occurs when the excitation is varied (Fig. 5.3.13). This suggests that the principal effect of detuning is a variation in gain and not the introduction of asymmetry. The non-symmetric behavior with respect to the gain line center is suggested to result from dispersion focusing (the refractivity of the laser medium changes sign at the gain line center. Since there is a radial structure in the active medium, it acts as a lens, positive on one side and negative on the other side of the line center).

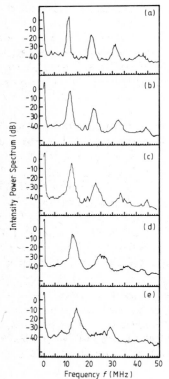

Fig. 5.3.12. The pulsation spectrum, which initially consists of lines, broadens with increasing discharge current. This is interpreted as an intermittency phenomenon. An increasing number of "intermittent" bursts would produce such spectra. The spectrum (e) clearly contains a broadband component indicative of chaos (Gioggia and Abraham, 1983a).

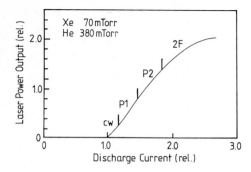

Fig. 5.3.13. Laser power output vs. excitation for the He-Xe ring laser (70 mTorr Xe, 380 mTorr He). Laser tuning near line center. The different pulsing regions are indicated (cw, continuous emission; P1, pulsing at frequency f; P2, pulsing at frequency $f/2$; $2F$, pulsing with two incommensaturate frequencies) (Tarroja et al., 1986).

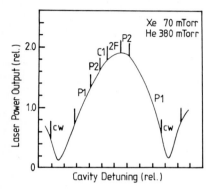

Fig. 5.3.14. Laser power output vs. laser resonator detuning with regions of different pulsing characteristic (C_1 chaotic, P_1 pulsing). Conditions as in Fig. 5.3.13 (Tarroja et al., 1986).

Tarroja et al. (1986) cautioned that the observed subharmonic (P2) must not really indicate a period doubling. In the intensity spectrum, the appearance of half of the pulsing frequency cannot be distinguished from a change from a symmetric to an antisymmetric pulsing (Fig. 5.3.4), because symmetric pulsing has only even harmonics in the spectrum, whereas antisymmetric pulsing has all harmonics. The transition from symmetric to antisymmetric pulsing therefore produces new spectral lines mid-way between those of the symmetric pulsing, like a period doubling of the pulsing frequency.

A distinction can only be made by measuring the field instead of the intensity. This is done by heterodyne detection, i.e. superposing the laser beam to be measured with the beam of another (stable) laser on a photon detector. The signal from the detector is then

$$(E_1 + E_2)^2 = E_1^2 + 2 E_1 E_2 + E_2^2$$

where E_i are the fields of the two lasers. In addition to the intensity spectrum E_1^2 and a DC component E_2^2, the interference term $E_1 E_2$ is proportional to E_1 and allows the optical spectrum to be measured. The experimental set-up used for the measurements is then more complex (Fig. 5.3.15).

The optical spectra for asymmetric pulsing have the strongest component at the laser frequency (the "carrier"), whereas this carrier is missing in the symmetric pulsing spectra.

Theoretically calculated optical spectra (Fig. 5.3.5) for a certain range of excitation of the He-Xe traveling wave laser reveal that the observed subharmonics (P2 regions) in the intensity spectra correspond to a symmetry change. At low excitation for central tuning of the ring laser the pulsing is asymmetric. A higher excitation then leads to chaotic spectra, after which a transition to a symmetric pulsing occurs. The corresponding measured spectra are shown in Fig. 5.3.16; (a) corresponds to stable operation of the test laser. The optical spectrum consists of only one frequency. In (b) pulsing in the intensity spectrum is shown. The optical spectrum signal-to-noise ratio is too low for the additional frequencies to be seen. Since there is only one carrier visible, however, it follows that the pulsing is asymmetric. In (c) and (d) chaotic emission recognizable in the intensity spectrum and the optical spectrum is shown and (e) and (f) show the emergence of an optical spectrum without a carrier, clearly indicating a symmetric pulsing.

Hence fair agreement between the model calculation and the measurements is found. It is noted that the chaotic regions become smaller as the laser resonator is tuned away from the gain line center. Figure 5.3.17 shows measurements taken at half Doppler-width detuning. The change from asymmetric to symmetric pulsing is still observed; however, the chaotic range has disappeared.

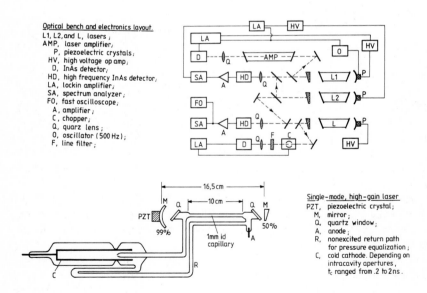

Fig. 5.3.15. Experimental set-up for heterodyne measurement of laser field spectra (Abraham et al., 1983).

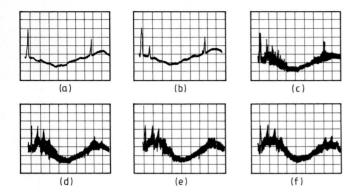

Fig. 5.3.16. Experimentally obtained heterodyne and homodyne spectra of pulsing He-Xe laser. The line at the left is a zero frequency marker. The lines on the right (clearly visible as a line in (a) and (b)) are the heterodyne signals. The laser intensity pulses (b) are indicated by the homodyne spectral line. The corresponding heterodyne spectral line(s) is (are) below the noise level (Tarroja et al., 1986).

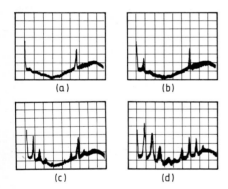

Fig. 5.3.17. Homodyne and heterodyne spectra of the detuned He-Xe laser. As in the resonantly tuned case (Fig. 5.3.16), a transition from asymmetric to symmetric attractor is observed when the excitation is increased (Tarroja et al., 1986).

Good agreement is found primarily in the most obvious quantity to measure, namely the pulsing frequency.

The theoretical and experimental results of Casperson et al. (1978) and Tarroja et al. (1986) differ because they are concerned with different parameter ranges. The lower pressure conditions in the former instance yield a more complex structure of the individual pulses.

The He-Xe system, in spite of the difficulty of carrying out theoretical modelling brought about by the Doppler broadening, is today the laser whose dynamics is most completely understood. It can therefore be used as a test object for refinements in theoretical modeling or for quantitative comparisons of measurements and model results in chaotic dynamics.

b. The 3.39 µm He-Ne laser

Experiments on a standing-wave He-Ne laser have also shown the occurrence of a laser instability (Gioggia and Abraham, 1983b). Various forms of self-pulsing have been observed: pulsing at a single frequency, simultaneous pulsing at two different frequencies, pulsing at a subharmonic of the original frequency and broad-band spectra underlying line spectra, which have been interpreted as chaotic dynamics. All these have been found near the laser Lamb-dip, which indicates that the standing-wave nature of the optical field plays an essential role. No theoretical treatment exists for modeling these experimental findings.

By using the heterodyne technique (mixing with a stable reference laser), it has been shown that the pulsing is of the symmetric type (central carrier missing)) (Abraham et al., 1983). Since no higher than the first subharmonic is observed, it appears plausible that this is again a symmetry change of the pulsing, as discussed for the He-Xe system. The fact that chaos sets in directly after the appearance of the subharmonic in the intensity spectrum supports this assumption. We recall that as in the He-Xe system it is found that chaos develops from the competition between the two pulsing forms.

6 Lasers with Saturable Absorbers

In a laser with a saturable absorber, two media are present inside the laser resonator. One of them is pumped so that the atoms have a positive population inversion (active or amplifying medium) and the other is left with a negative population inversion (passive or absorbing medium). The laser field may saturate the absorption, or the field and the absorber may interact in other ways so that time-dependent or bistable laser behavior can result. Lasers with saturable absorbers are primarily used to obtain pulsed emission with short and intense repetitive pulses ("passive Q-switching"). Interest has recently expanded because a large variety of unstable behaviors were observed which are currently being studied from the more general aspect of non-linear dynamics in optical systems.

Typical examples of lasers with saturable absorbers are CO_2 gas lasers and semiconductor lasers. In this chapter we study the dynamic behavior with single-mode emission.

6.1 Theoretical Description

Almost all theoretical analyses have considered the case of homogeneously broadened single-mode uni-directional lasers, in order to avoid complexity brought about by inhomogenous broadening and standing-wave effects. We first treat the case of mid-infrared gas lasers and then semiconductor lasers.

6.1.1 CO_2 (and similar) Gas Lasers

Several theoretical models have been developed for the description of CO_2 lasers (or N_2O or CO lasers) with saturable absorbers such as CH_3I, CH_3F, SF_6, CH_3OH or HCOOH, each involving different kinds of physical or mathematical approximations.

Models

An early simple model considered the active medium as a two-level system in which the atomic polarization can be adiabatically eliminated (see Section 3.1.2.c) from eqns. (5.1.19), because for CO_2 lasers the inequality $\gamma_\parallel < \gamma_\perp, \kappa$ is usually fulfilled, and the action of the passive medium was simply accounted for by replacing the constant field decay rate κ by an intensity-dependent rate $\kappa(E)$ given by

$$\kappa(E) = \kappa_0 + \frac{\kappa_s}{1 + |E|^2/I_s} \tag{6.1.1}$$

where κ_0, κ_s and I_s (resonator loss, linear absorption of saturable absorber and its saturation intensity, respectively) are constants depending on the choice of the absorbing medium (Haken, 1985). The decrease in $\kappa(E)$ when E increases describes the saturation of the absorption in the passive medium when the laser intensity increases.

An improved model, known as the two-level model, views the amplifying and absorbing media as independent two-level systems coupled to the same laser field (Burak et al., 1971; Lugiato et al., 1978; Antoranz et al., 1982; Mandel and Erneux, 1984; Chyba et al., 1987; Chyba, 1988). Hence eq. (5.1.19) can be generalized to the following set:

$$\dot{E}(t) = -(\mathrm{i}\omega_c + \kappa)E(t) + Ng\,p(t) + \bar{N}\bar{g}\,\bar{p}(t)$$

$$\dot{p}(t) = -(\mathrm{i}\omega_{ab} + \gamma_\perp)p(t) + g\,E(t)d(t)$$

$$\dot{\bar{p}}(t) = -(\mathrm{i}\bar{\omega}_{ab} + \bar{\gamma}_\perp)\bar{p}(t) + \bar{g}\,E(t)\bar{d}(t) \tag{6.1.2}$$

$$\dot{d}(t) = -\gamma_\parallel\,[d(t) - d_0] - 2g\,[E(t)p(t)^* + \text{c.c.}]$$

$$\dot{\bar{d}}(t) = -\bar{\gamma}_\parallel\,[\bar{d}(t) - \bar{d}_0] - 2\bar{g}\,[E(t)\bar{p}(t)^* + \text{c.c.}]$$

where the quantities with bars refer to the absorbing medium and those without bars correspond, as before, to the amplifying medium. In the first equation a term has been added to take into account the field absorption by the absorbing medium.

The set (6.1.2) consists of five complex or eight real coupled equations (E, p and $\bar{p}$ are in general complex quantities). Except in Chyba et al. (1987) and Chyba (1988), they have been solved only in the resonant case, i.e. when $\omega_c = \omega_{ab} = \bar{\omega}_{ab}$. As was shown in Section 5.2 with normal lasers without a saturable absorber, this implies that the phases in $E(t)$, $p(t)$ and $\bar{p}(t)$ are equal (except for the fast oscillating factor $\exp(-\mathrm{i}\omega_L t)$); see eqns. (5.2.2)−(5.2.6); however, in spite of this fact in the present case the phases can greatly influence the results under certain conditions (Antoranz et al., 1982; Mandel and Erneux, 1984), and may have an influence leading in particular to shifts in the laser frequency (i.e. $\omega_L \neq \omega_c$) and to a dependence on the symmetry of the initial conditions (Mandel and Erneux, 1984). Nonetheless, the phases are often ignored as well. This allows eqns. (6.1.2) to be reduced to a set of five real equations.

In many cases $\gamma_\perp$ and $\bar{\gamma}_\perp$ are much larger than the remaining relaxation rates k, $\gamma_\parallel$ and $\bar{\gamma}_\parallel$, so that the molecular polarizations $p(t)$ and $\bar{p}(t)$ can be adiabatically eliminated (Section 3.1.2.c), i.e. they can be calculated as a function of the instantaneous value of the field and population inversions and the equations for $p(t)$ and $\bar{p}(t)$ disappear from eqns. (6.1.2). This approximation leads to a rate-equation model with only three coupled real differential equations. Mandel and Erneux (1984) made a comparison between the models with eight, five and three equations for a particular case. The validity of the three-equations model is discussed in Mandel and Erneux (1984) and in Arimondo et al. (1983). In Mandel and Erneux (1984) it is reported that, for certain typical values of the relaxation rates and of $g/\bar{g}$, the sets of eight and five coupled equations lead to stable periodic solutions and in particular to passive Q-switching, whereas for the set of three equations the only stable solutions are steady states.

This two-level model is useful for rigorously describing the direct interaction phenomena between the field and the resonant molecular transition of the active and passive media, but is too crude to describe properly the multiplicity of vibrational and rotational levels present in the molecules interacting directly or indirectly with the field. To take other energy levels into account at least in an approximation, the so-called four-level model (Burak et al., 1971; Arimondo et al., 1983; Dupré et al., 1975) considers both the active and passive media as four-level systems (Fig. 6.1.1). A pair of these levels, j_1 and j_2 (let us consider, for instance, the active medium), is the one with which the field is resonantly coupled, and the other pair of levels, 1 and 2, represent all the other rotational levels belonging to the same vibrational band as j_1 and j_2. The level j_1 is coupled to the set of rotational levels 1 (and j_2 is coupled to the set 2) via rotational-relaxation collision processes described by the transition rates γ_R and γ_R'. The vibrational bands 1 and 2 are in turn coupled to any other band through rotational-vibrational relaxation processes described by the respective rates γ_1 and γ_2. A similar scheme applies to the passive medium. The pumping mechanism, described by a constant pumping rate, is assumed to act on the vibrational band 2 of the active medium.

With the four-level model, the time evolution of the populations of the levels, $M_{j1}(t)$, $M_{j2}(t)$, $M_1(t)$, $M_2(t)$ (and $\bar{M}_{j1}(t)$, $\bar{M}_{j2}(t)$, $\bar{M}_1(t)$, $\bar{M}_2(t)$ for the passive medium), is better described than with the two-level model, because the relaxation rates γ_R, γ_R' are usually two or three orders of magnitude larger than γ_1, γ_2. If rate equation conditions are assumed (i.e., the polarization and field-phase effects are ignored), the four-level model leads to a system of nine coupled differential equations (one for the photon number

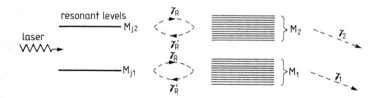

Fig. 6.1.1. Scheme of the four-level model for the amplifying medium. The resonant levels are j_1, j_2 (with populations M_{j1}, M_{j2}) and the other close rotational levels are 2, 1 (with populations M_2, M_1). The population-transfer collision processes between the levels j_1 and 1 (and also between j_2 and 2) are described by the transition rates γ_R, γ_R', and the decay from the levels 1, 2 to other ro-vibrational levels are described by the respective rates γ_1, γ_2. A similar picture applies to the absorbing medium. (From Arimondo et al., 1983.)

density and eight for the populations of the involved levels). If it is assumed that $\gamma_1 = \gamma_2$ and $\bar{\gamma}_1 = \bar{\gamma}_2$ (in spite of the fact that usually one has $\gamma_2 < \gamma_1$ and $\bar{\gamma}_1 \simeq \bar{\gamma}_2$), one can work directly with the population inversions $D_j = M_{j2} - M_{j1}$, $D = M_2 - M_1$, $\bar{D}_j = \bar{M}_{j2} - \bar{M}_{j1}$ and $\bar{D} = \bar{M}_2 - \bar{M}_1$ and the system reduces to five equations. Since in general $\gamma_R, \gamma'_R \gg \gamma_1, \gamma_2$, one can adiabatically eliminate the variables D_j and $\bar{D}_j$ and the system finally reduces to three coupled equations.

The three-equation, four-level model has been applied to several cases (Arimondo et al., 1983 and 1985), yielding results that are in good qualitative and partially quantitative agreement with the experimental findings.

Ueda and Shimizu (1984) and Tachikawa and co-workers (1986 and 1987) considered that vibrational relaxation of the lower laser level, which is not taken into account well in the two- and four-level models, may strongly influence the dynamics of passive Q-switching, because its characteristic time is of the same order of magnitude as the typical undamped pulsations that are experimentally observed. The rotational relaxation in the gain medium is too fast (two or three orders of magnitude) to play a dominant role in the time-dependent behavior. On the other hand, since in the absorbing medium the pressure is usually two orders of magnitude smaller than in the gain medium, there the rotational relaxation plays the dominant role instead of the vibrational relaxation. They therefore proposed a relatively simple model known as the three-level-two-level model (Tachikawa et al., 1986), which seems to reproduce their experimental results excellently (Tachikawa et al., 1986, 1987 and 1988; Tachikawa, Hong et al., 1988a and b; Tannii et al., 1988). In this model, the amplifying medium is regarded as a three-level system and the absorbing medium is considered as a two-level system (Fig. 6.1.2). With the notation indicated in the figure caption, and within the rate-equation approximation, the dynamical equations are

$$\dot{n}(t) = B_g n (M_1 - M_2) l_g / L - B_a n N l_a / L + A M_1 - 2\kappa n$$
$$\dot{M}_0(t) = R_{20} M_2 + R_{10} M_1 - (P + R_{02}) M_0$$
$$\dot{M}_1(t) = -B_g n (M_1 - M_2) + P M_0 + R_{21} M_2 - (R_{10} + R_{12}) M_1 \qquad (6.1.3)$$
$$\dot{M}_2(t) = B_g n (M_1 - M_2) + R_{12} M_1 + R_{02} M_0 - (R_{21} + R_{20}) M_2$$
$$\dot{N}(t) = -2 B_a n N - r(N - N^*)$$

where the coefficients B_g and B_a are the cross-sections multiplied by the light velocity c for the induced emission in the gain medium and the absorption in the absorbing medium, respectively; 2κ represents, as before, the resonator intensity loss rate; L, l_g and l_a denote the length of the laser resonator, the gain tube and the absorption cell, respectively; $N(t) = N_0(t) - N_1(t)$ and N^* represents the thermal equilibrium value of $N(t)$. The coefficients B_g and B_a are given by

$$B = \frac{2}{3} \frac{\pi \cdot \mu^2 \omega_L}{3 \varepsilon_0 l_a \Delta\omega} \qquad (6.1.4)$$

where μ is the transition dipole moment, ω_L the laser frequency and $\Delta\omega$ the Doppler half-width. The rate A of spontaneous emission from the upper laser level is usually neglected, and the upward relaxation rates R_{02} and R_{21} can also be neglected in general.

In cases where the pressure of the absorbing medium is higher (Tanii et al., 1988), a Voigt profile was adopted for the absorption line shape, which allows a better estimate of the actual value of B_a by taking into account its frequency dependence (Tanii et al., 1988).

Because the three-level system adopted for the gain medium represents a closed system (Fig. 6.1.2), the total population of the vibrational levels $M = M_0 + M_1 + M_2$ remains constant, so that eqns. (6.1.3) represent a system of four independent coupled equations, which in principle allows for pulsed and chaotic dynamics (Section 3.3.2.b).

Very recently a more improved model (Tomasi et al., 1989 and Ref. 17 therein) and a simplified version (Dangoisse and Glorieux, to be publ.) have been given. In the paper by Tomasi et al. (1989) a discussion of the laser parameters used in Tachikawa et al. (1986, 1987 and 1988), Tannii et al. (1988) and Tachikawa, Hong et al. (1988a, and b) is included.

Results

From the different results obtained from the models described above and their comparison with experimental results* (to be considered in Section 6.2), one can draw the following conclusions about the dynamic behavior of gas lasers with a saturable absorber:

(i) There exist three steady-state solutions: the trivial zero-intensity solution $I_0 = 0$, which is stable only for low pumping rates, and two non-zero intensity solutions

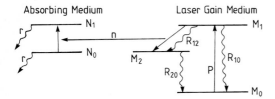

Fig. 6.1.2. Three-level (gain medium)-two-level (absorbing medium) rate equations model for describing passive Q-switching. $M_i = (i = 0, 1, 2)$ is the population density of the vibrational level i, N_j ($j = 0, 1$) is the population density of the rotational level j, n is the photon density, R_{ij} ($i,j = 0, 1, 2$) represents the vibrational relaxation rate from level i to level j in the amplifying medium, r is the rotational relaxation rate in the absorbing gas and P denotes the pumping rate. (From Tachikawa et al., 1986.)

* Burak et al., 1971; Salomaa and Stenholm, 1973; Dupré et al., 1975; Dembinski et al., 1978; Lugiato et al., 1978; Mrugala and Peplowski, 1980; Erneux and Mandel, 1981; Antoranz et al., 1982; Jacques and Glorieux, 1982; Arimondo et al., 1983; Heppner et al., 1984; Mandel and Erneux, 1984; Ueda and Shimizu, 1984; Arimondo et al., 1985; Tachikawa et al., 1986; Velarde and Antoranz, 1986; Chyba et al., 1987; Tachikawa et al., 1987; Tachikawa et al., 1988; Tachikawa, Hong et al. 1988a and b; Tanii et al., 1988; Tomasi et al., 1989; Dangoisse and Glorieux, to be published.

I_+ and I_-, which appear beyond a threshold pumping rate and under certain conditions may behave as in Fig. 6.1.3. The solution I_0 is stable for $0 \leqslant A \leqslant A_0$ (where A represents a pump parameter that is proportional to the pumping rate and to the induced emission coefficient in the amplifying medium); the solution I_- is always unstable, and the solution I_+ can be either stable for any value of $A \geqslant A_1$ (as in the case depicted in Fig. 6.1.3), or it may possess a domain of instability in the lower part of the branch (i.e. for low values of A ranging from A_1 in Fig. 6.1.3 to some higher value A_2). Depending on the relative values of A_0 and A_2, we may have domains of stable steady-state emission (for instance, when only the I_+ solution is stable), bistability between I_0 and I_+ steady-state solutions (as in the case in Fig. 6.1.3), or time-dependent emission (when both I_0 and I_+ are unstable), which may be related to passive Q-switching. The possibilities of the simultaneous or (more frequently) exclusive appearance of bistability and passive Q-switching are analysed in the papers by Arimondo et al. (1983), Tachikawa et al. (1987), Jacques and Glorieux (1982), Heppner et al. (1984), Dembinski et al. (1978) and Erneux and Mandel (1981). The appearance of hysteresis cycles was predicted by Salomaa and Stenholm (1973).

(ii) Different kinds of time-dependent behavior have been predicted, most of which are represented by those shown in Fig. 6.1.4, where (a′) (which shows high-intensity peaks), (c′) and (d′) show the most typical passive Q-switching behaviors (let us call them type I passive Q-switching), and (b′) corresponds to a sinusoidally modulated laser output which constitutes a special kind of passive Q-switching that will be called type II. These periodic behaviors usually appear for different values of the control parameters, but overlappings between their domains of existence (and those of the steady-state solutions) can occur, giving rise to generalized bistability (or "bi-rhythmicity") (Section 3.3.3.c) (Mandel and Erneux, 1984; Arimondo et al., 1985). All features in (a′)−(d′) except for the undamped undulation in (c′) have been correctly predicted by the three models described above. The undamped undulations have been better reproduced by the three-level-two-level model (Tach-

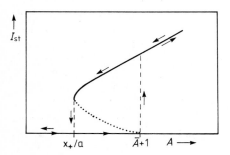

Fig. 6.1.3. Intensity of the steady-state solutions versus the pump parameter A, as predicted by the four-level model. Continuous line, I_+ solution; dotted line, I_- unstable solution; horizontal axis, $I_0 = 0$ solution. Bistability between I_+ and I_0 is possible, with a hysteresis cycle indicated by arrows. In the main text the limits x_+/a and $\bar{A} + 1$ of the bistable domain are denoted by A_1 and A_0, respectively. (From Arimondo et al., 1983.)

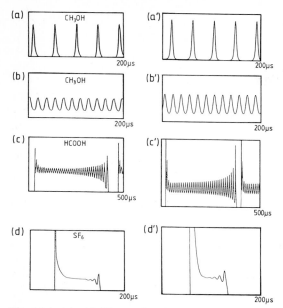

Fig. 6.1.4. (a)–(d) Observed and (a')–(d') calculated pulse shapes, showing the different dynamic behaviors found for lasers with a saturable absorber. The observed behavior is commented on in Section 6.2.1. The numerically obtained behaviors are predicted by the three-level-two-level model of eqns. (6.1.3). (From Tachikawa, Tanii and Shimizu, 1988.)

ikawa et al., 1986 and 1987), which attributes them essentially to the competition between the rate of induced emission from the upper laser level and the rate of vibrational relaxation from the lower laser level.

The type II passive Q-switching may appear from the I_+ steady state, through a supercritical Hopf bifurcation when the pump parameter is adiabatically increased (Arimondo et al., 1985; Tanii et al., 1988). It has also been predicted as a Hopf bifurcation of the I_0 state (Mandel and Erneux, 1984), but this has not been observed (Arimondo et al., 1983). The mean intensity and/or the modulation depth may increase strongly when the pump parameter is further increased. The type I passive Q-switching behavior may arise through a subcritical Hopf bifurcation from the I_+ state (hard-mode excitation). Phase diagrams showing the domains of existence of steady-state emission, sinusoidal modulation and passive Q-switching have been calculated (Tanii et al., 1988; Tachikawa et al., 1988).

(iii) Chaotic behavior has also been found in lasers with saturable absorbers. The efforts devoted to the interpretation of chaotic behavior have resulted in a better understanding of passive Q-switching and a fairly global picture for all the regimes has been obtained**.

** Mrugala and Peplowski, 1980; Antoranz et al., 1982; Velarde and Antoranz, 1986; Antoranz and Rubio, 1988; Tachikawa et al., 1988; Tachikawa, Hong et al., 1988 a and b; Arimondo et al., 1988; Bekkali et al., 1988; Dangoisse et al., 1988; Erneux, 1988; Hennequin et al., 1988; Tomasi et al., 1989; Dangoisse and Glorieux, to be published.

Let us consider the most interesting conditions where both steady-state solutions I_0 and I_+ are unstable. With increase in the pump parameter, a limit cycle arises from the fixed point associated with the solution I_+ through the above-mentioned super-critical Hopf bifurcation and progressively increases in size (Fig. 6.1.5.b). This is the type II passive Q-switching.

By varying some control parameter, the orbit in the phase space may change con-siderably and come close to a homoclinic orbit (Section 3.3.3.b) connected to the saddle point (or saddle node) I_0. As in such a kind of orbit the system remains close to the saddle point for a long period, this explains the type I behavior in Fig. 6.1.4.a': the output intensity remains close to zero for long intervals, separated by short and intense peaks corresponding to the fast ejection from the I_0 point and to the return to it. This is illustrated by the trajectory in Fig. 6.1.5.a, where one can observe that in fact in the zero-intensity periods the orbit remains in the $I = 0$ plane but does not necessarily come close to the I_0 point.

If the pump parameter (or some other parameter) is increased, the homoclinic orbit connected to the I_0 point progressively approaches the saddle focus I_+. It approaches I_+ along the stable manifold and spirals outward on the unstable manifold (Arimondo et al., 1988; Bekkali et al., 1988; Dangoisse et al., 1988; Hennequin et al., 1988; Tomasi et al., 1989; Dangoisse and Glorieux, to be published) (see Fig. 6.2.4). Hence the dy-namics around the saddle focus I_+ are of the Shilnikov type (Section 3.3.3.b) but the dynamical evolution occurs in the opposite direction ("inverse Shilnikov" dynamics (Dangoisse et al., 1988)). The condition for Shilnikov chaos referred to the relative values of the characteristic exponents is also the inverse of that of the direct case in Fig. 3.3.5. The fact that the orbit remains connected to the I_0 point introduces a further difference from the pure Shilnikov case, because the whole orbit is a homoclinic cycle instead of a simple homoclinic orbit. This affects the kind of chaotic behavior that can be found for the trajectories close to the I_+ point, reducing the possibilities of its appearance (Arimondo et al., 1988; Hennequin et al., 1988; Tomasi et al., 1989). The inverse Shilnikov dynamics explain the kind of behavior in Fig. 6.1.4.c' or d': the large

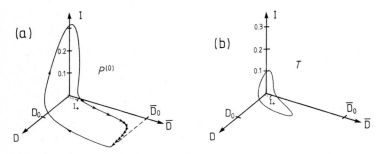

Fig. 6.1.5. Numerically calculated orbits in the three-dimensional phase space $(D, \bar{D}, I)$, where D and $\bar{D}$ are the population differences between the amplifier vibrational levels and the absorber rotational levels, respectively, and I is the laser intensity. The pumping rate is 1.85 times larger in (b) than in (a). (a) and (b) are examples of type I and type II passive Q-switching, respectively. The zero intensity solution I_0 has the coordinates $(D_0, \bar{D}_0, 0)$. (From Tomasi et al., 1989.)

initial peak corresponds to the trajectory far from the I_+ point, whereas the undulations are due to the outward spiraling around I_+. Indeed, by varying a control parameter in such a way that the homoclinic orbit progressively approaches the I_+ point and progressively increases the number of spiral turns around it, we obtain a sequence of behaviors going from that in Fig. 6.1.4.a′ (i.e. zero undulation after the main peak) to that in Fig. 6.1.4.c′ with an increasing number of undulations after the main peak. This has been analysed with the three-level-two-level model (with some improvements) (Arimondo et al., 1988; Bekkali et al., 1988; Dangoisse et al., 1988; Hennequin et al., 1988) and, as shown below, has also been experimentally observed in detail. In particular, the experimental results in Fig. 6.2.2 have been well reproduced numerically (Bekkali et al., 1988).

Chaos has been found from either type I or type II instabilities through period-doubling sequences (Arimondo et al., 1988; Bekkali et al., 1988; Dangoisse et al., 1988; Hennequin et al., 1988; Tomasi et al., 1989; Dangoisse and Glorieux, to be published). In a different approach, Tachikawa, Tanii and Shimuzu (1988), found chaos (which has recently been observed (Tachikawa, Hong et al., 1988a and b)) in the region of overlap between type I and type II pulsing in the phase diagram; chaos also appeared through a period-doubling sequence.

Generalized bistability (Section 3.3.3.c) between cw regime I_+ and type-I instability, "off" regime I_0 and type II instability, I_+ regime and type II instability, and type I and type II instabilities is possible (Arimondo et al., 1988; Tomasi et al., 1989).

As a final comment let us mention the work of Garcia-Fernández and Velarde (1988), who took into account space inhomogeneity and detuning effects within the two-level model (space dependence increases the domain of the I_0 state, thus delaying the laser action); Antoranz and Rubio (1988), who predicted hyperchaos (in the sense that an attractor has simultaneously two positive Lyapunov exponents; Section 3.3.3.c), also with the two-level model, near a codimension-two bifurcation point, which presumably will be difficult to reach experimentally; and Englund (1988), who described Lamb-dip effects appearing when a standing-wave laser field is generated.

6.1.2 Semiconductor Lasers

Semiconductor lasers divided into two (or more) sections along the laser cavity have been built, in which one of the sections acts as the amplifying medium and the other as the absorbing medium. Hence they constitute simple and compact generators for pulsed optical radiation.

Because of electron-phonon and electron-impurity interactions, the polarization relaxation is much faster than carrier population relaxation, so that a rate-equation model taking as variables the number of carriers n_e and $\bar{n}_e$ in the amplifying and absorbing sections, respectively, and the photon number n_p is usually considered. Depending on the way in which the amplifying and absorbing sections are created, the set of rate equations may take different forms. As an example, let us consider the case where these sections are distinguished by different drive currents I and $\bar{I}$ and by different non-radiative carrier lifetimes τ_e and $\bar{\tau}_e$, in which the equations take the following form (Kawaguchi and Iwane, 1981; Kawaguchi, 1982 and 1984):

$$\frac{dn_p}{dt} = -n_p/\tau_p + n_p v_g [\xi g(n_e) + \bar{\xi} \bar{g}(\bar{n}_e)] + \beta B(\xi n_e^2 + \bar{\xi} \bar{n}_e^2)$$

$$\frac{dn_e}{dt} = P - n_e/\tau_e - g(n_e)n_p v_g - Bn_e^2 \qquad (6.1.5)$$

$$\frac{d\bar{n}_e}{dt} = \bar{P} - \bar{n}_e/\bar{\tau}_e - \bar{g}(\bar{n}_e)n_p v_g - B\bar{n}_e^2$$

where n_p, n_e and $\bar{n}_e$ are defined per unit volume, $P = I/qV$ is the pump rate per unit volume (I is the injection current, q is the carrier charge and V is the volume), B is the recombination coefficient for carriers, v_g is the group velocity of light in the medium, $g(n_e) = an_e + b$ and $\bar{g}(\bar{n}_e) = a\bar{n}_e + b$ represent the gain (or losses, if it takes negative values) in each section, $\xi = l/(l + \bar{l})$ and $\bar{\xi} = \bar{l}/(\bar{l} + l)$, where l and $\bar{l}$ are the lengths of the amplifying and absorbing sections, respectively, and β is the spontaneous emission coefficient. In the simpler conventional rate equations the terms with the recombination coefficient B are neglected (see, for instance, eqns. (7.1.16)).

Because there are three equations in the set (6.1.5), chaotic emission is in principle possible. However, in the work of Kawaguchi (1984), pulsing emission was searched, which could be useful for communication applications. For typical values of the parameters, the phase diagram in Fig. 6.1.6 was obtained, in which a region of pulsing is predicted for $\bar{\tau} < 1$ ns ($\tau = 20$ ns was considered). Further, a bistable behavior appears for $\bar{\tau} > 2$ ns. Kawaguchi (1984) also considered the influence of a sinusoidal modulation of the pump rate, which is described in Chapter 7.

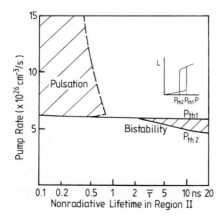

Fig. 6.1.6: Phase diagram in the $(\bar{\tau}, P)$ plane, showing the threshold pump rates (continuous line) and the domains of pulsing emission and bistability (shaded regions). The inset shows the P_1/L curve in the bistable domain ($P_1 = P$ in our notation). (From Kawaguchi, 1984.)

6.2 Experiments

6.2.1 CO$_2$ Laser

Experiments were conducted with a CO$_2$ laser operating on the P(32) line in the 10 μm wavelength range. Gaseous CH$_3$I was used as the saturable absorber at low pressure (Dangoisse et al., 1988).

For very low pressure (ca. 40 mTorr) the laser operates continuously and for high pressure (ca. 500 mTorr) the laser is off. In the range between these two pressures, passive Q-switching is obtained which results in a modulated output. The modulation can vary from weak sinusoidal modulation (type II Q-switching) (Fig. 6.1.4(b)) to the emission of short (<100 ns) giant pulses with peak power up to 100 times the cw power of 1 − 100 kHz repetition frequency (type I Q-switching) (Fig. 6.1.4(a)).

The passive Q-switch CO$_2$ laser has been studied experimentally in many investigations, mostly with the aim of applying the giant pulses obtained. The pulse shape often observed is shown in Fig. 6.2.1(b) (see also Figs. 6.1.4(c) and (d)). The laser emission starts with a giant pulse, after which the intensity settles to almost a cw value. Growing oscillations around the cw value develop until their intensity touches zero, where it remains until inversion has built up sufficiently for the next giant pulse, and so on.

In the reported experiment (Dangoisse et al., 1988), the CO$_2$ laser is tuned first to the gain line center and the absorber pressure is adjusted to yield a small-amplitude modulation of the laser output (Fig. 6.2.2(a)). Detuning of the laser results in deeper modulation, which eventually period doubles (Fig. 6.2.2(b)) and quadruples (Fig. 6.2.2(c)). After the period-doubling cascade, chaos is reached (Fig. 6.2.2(d)). After the chaotic range, ranges are found in which a large pulse is followed by one, two, three or four oscillations. Fig. 6.2.2(e) shows the case of a large pulse followed by two oscillations ($T\,12$) and Fig. 6.2.2 (h) the large pulse followed by one oscillation ($T\,11$). Between the T_{11} and T_{12} range a chaotic range of irregular pulsing occurs (Fig. 6.2.2(f)).

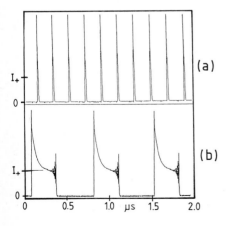

Fig. 6.2.1. Pulses of CO$_2$ laser with saturable absorber. (a) Homoclinic orbit around I_0; (b) homoclinic cycle connecting $I_0 + I_+$. (From Hennequin et al., 1988.)

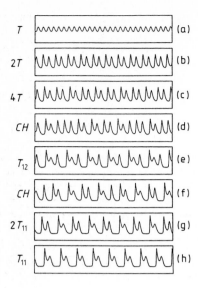

Fig. 6.2.2. Passive Q-switching pulse shapes as function of laser resonator detuning. The small-amplitude orbit period doubles to chaos followed by homoclinic cycles of the Shilnikov type, periodic or chaotic. For details see text. (From Dangoisse et al., 1988.)

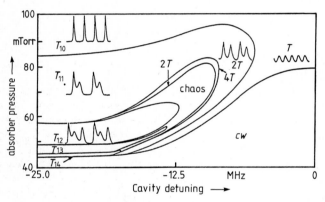

Fig. 6.2.3. State diagram of the CO_2 laser with a saturable absorber with the pulse shapes of Fig. 6.2.2. Note the connections between different pulse shapes, e.g. $2T \rightarrow T_{11}$. (From Dangoisse et al., 1988.)

The chaotic dynamics brought about by competition of the different temporal patterns is reminiscent of the case of the inhomogeneously broadened laser (see Section 5.3), where chaos occurs through the competition of the symmetric and asymmetric pulsing.

By choosing particular conditions, pulses followed by more than two oscillations (T_{13}, T_{14} ..., and in general pulses like Fig. 6.2.1(b)) can be found. They are found to period double to their respective chaotic ranges. The ranges of the different pulse shapes are given in the phase diagram in Fig. 6.2.3. At the limits the pulse shapes change abruptly (instability boundaries).

Period-doubling cascades to chaos of the individual pulse types T_{1i} are generally observed. For example, Fig. 6.2.2(g) shows the first subharmonic of T_{12}.

The experimental findings can be explained by the simple attractor structure described previously (Section 6.1.1). The dynamics of the system is dominated by two unstable fixed points: the zero intensity point, which is a saddle point, and the steady-state fixed point, which is a saddle focus (Fig. 6.2.4).

Starting at the steady-state value, the system spirals away from the unstable focus point until it touches the zero intensity point. After sufficient inversion has built up (since no laser emission occurs), the system is ejected in a giant pulse (very similar to the generation of the giant pulses in the laser with injected signals, Section 7.3) from the zero intensity point and subsequently falls back into the steady-state focus point from which it spirals out again, and so on. The different pulse shapes observed in the experiment differ in the number of orbits around the focus fixed point that are needed to bring the system to the zero intensity point; In T_{11} it is one, in T_{12} two, and so on.

This kind of behavior is typical of a Shilnikov orbit (Shilnikov, 1965), as described in Chapter 3. Differently, however in the present case as mentioned in Section 6.1.1, the trajectory passes close to two fixed points instead of only one, so that it reveals the existence of a homoclinic cycle instead of a homoclinic Shilnikov-type orbit. The laser with a saturable absorber can be called "inverse Shilnikov", since the focus is repelling and the dimension perpendicular to the spiral plane attracting (time-reversed Shilnikov dynamics). Consequently, chaos in this system can occur if the positive Lyapunov exponent associated with the dimension perpendicular to the spiral plane exceeds in magnitude the real part of the complex conjugate Lyapunov exponents of the focus.

The small amplitude modulation initially mentioned (Fig. 6.2.2(a)) comes about by a supercritical Hopf bifurcation of the steady-state emission fixed point, after which the system moves in a limit cycle around the steady-state point with a revolution frequency corresponding to the spiral frequency. The laser detuning then leads to period doubling, chaotic emission and finally to the repulsion of the steady-state fixed point. The last effect results in the outward spiraling around the steady-state emission point and the eventual capture of the system by the zero intensity saddle point.

The "normal" Q-switch pulses (Fig. 6.2.1(a)) correspond to conditions where the steady-state emission point is attractive enough for the system, after being ejected from

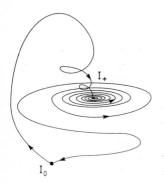

Fig. 6.2.4. Homoclinic cycle representation in phase space connecting two unstable (saddle) points. This Shilnikov geometry underlies the dynamics of, e.g., the laser with a saturable absorber. (From Hennequin et al., 1988.)

the zero intensity saddle point, to execute one large loop around the steady-state emission point and then fall back onto the zero intensity point, thereby emitting one regular giant Q-switch pulse (see also Fig. 6.1.5a).

6.2.2 Modulated Lasers

When a laser with a saturable absorber is operated in the continuous regime but close to the onset of Q-switch pulsing, its resonance (relaxation) frequency becomes very weakly damped. The system can then be very easily destabilized by modulation of a system parameter.

A corresponding experiment was carried out using an N_2O laser with a an ammonia gas cell as the saturable absorber (Hennequin et al., personal communication). The P(13) line of the N_2O laser in the 10 μm range was used, which has a precise spectral coincidence with the aQ(8,7) absorption line of $^{14}NH_3$. $^{14}NH_3$ has a relatively strong Stark effect in the ground state. The parameter modulated was therefore an electric field across the ammonia cell, which led effectively to a modulation of the absorber population.

The laser emission pulses obtained bear some resemblance to the Q-switched pulses obtained on Q-switching without modulation (Fig. 6.2.5). The emission starts with a giant pulse, which, however, in contrast to regular Q-switching, is followed by damped relaxation oscillation. The laser finally reaches the unstable steady-state emission level from which it spirals out as in the regular Q-switching until the zero intensity point is touched.

Apart from the damped oscillations following the first giant pulse, the modulated laser with a saturable absorber is different from typical Q-switching in that the number of oscillation pulses varies randomly from one period to the next, apparently indicating chaotic dynamics.

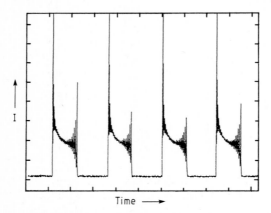

Fig. 6.2.5. Example of pulse shape emitted by a laser with a modulated absorber. Note that the pulse shape is chaotic, i.e. irregular. (From Hennequin et al., personal communication.)

Since the phase space of this system is four-dimensional, chaotic dynamics are in fact not unexpected. The observations are qualitatively fairly well reproduced by model calculations (Hennequin et al., personal communication). Figure 6.2.6 shows the calculated pulse shape for the passive Q-switching laser and the modulated laser with a saturable absorber. The latter clearly shows the irregularity of the pulses. No pictorial interpretation of the dynamics of this system in terms of unstable foci and unstable orbits has been given so far.

6.2.3 Semiconductor Diode Lasers

A similar behavior is observed with self-pulsing diode lasers (Kusnetzow et al., 1986). Self-pulsing in diode lasers is achieved by introducing saturable absorbers into the active medium. This can be done by doping or also simply by controlled optical damage (Winful et al., 1986). With a saturable absorber the laser also shows passive Q-switching which produces a train of pulses. Since all rates of the semiconductor lasers are larger than for gas lasers, the pulsing time scale is considerably shorter. Pulse widths are in the nano- to picosecond range and the pulse repetition frequencies are typically in the gigahertz range.

For these self-pulsing lasers, a transition to chaos via period doubling has been calculated when the current is modulated (Kawaguchi, 1984). In fact, period doubling and chaotic dynamics have been observed (Kawaguchi, 1984) (Fig. 6.2.7). Another experiment (Winful et al., 1986) has shown that period doubling for this system is an exception. As one would guess, with the system showing two frequencies, a quasi-periodicity transition to chaos is more likely.

The self-pulsing frequency is evidently influenced by the modulation. In particular, the self-pulsing frequency can lock with the modulation frequency, which means that the self-pulsing frequency can be controlled by the modulation frequency. In the case of locking, the ratio of the two frequencies becomes constant at a rational number.

(a)

(b)

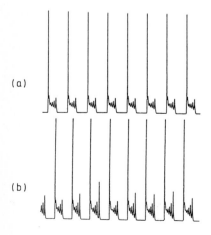

Fig. 6.2.6. Numerical calculation of pulses emitted by a laser with a saturable absorber (a) and a laser with a modulated saturable absorber (experiment) (b). (From Hennequin et al., personal communication.)

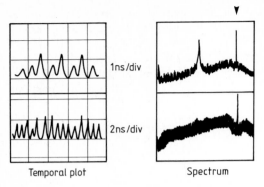

Temporal plot   Spectrum

Fig. 6.2.7. Observed pulse shapes (left) and intensity spectra (right) of a modulated laser diode that contains a saturable absorber. The sharp line is the modulation frequency (1 GHz). Top: Period-two pulsing; bottom: chaotic pulsing. (From Kawaguchi, 1984.)

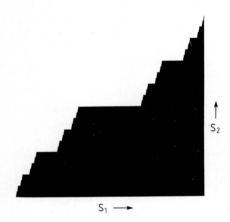

S_2

$S_1 \longrightarrow$

Fig. 6.2.8. "Devil's staircase" showing self-similarity: any part of the staircase magnified is identical with the whole. This figure is the integral over the Cantor set (Fig. 2.1.6a).

The frequency range over which the self-pulsing frequency is locked to the modulation frequency decreases the more "irrational" the ratio of the two frequencies is and increases with increasing modulation strength. The locking ranges for a given modulation form a "devil's staircase" (Fig. 6.2.8), the integral over the well known Cantor set, and thus form a fractal set.

The locked regions in a plane of modulation frequency vs. modulation amplitude form what is called "Arnold's tongues". Figure 6.2.9 shows that the tongues rapidly become smaller as the nominator and denominator increase, that is, as the "irrationality" increases. An attempt was made (Winful et al., 1986) to prevent locking by simultaneously changing the modulation amplitude and modulation frequency. The frequency ratio was kept at the two "very irrational" values $(\sqrt{5} - 1)/2$ (the "golden mean") and $\sqrt{2} - 1$ (the "silver mean"). By keeping these irrational ratios while increasing the modulation depth, it was possible to demonstrate the quasi-periodic transition to chaos.

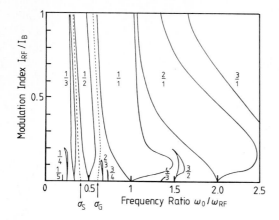

Fig. 6.2.9. Frequency-locked regions for a modulated semiconductor laser with a saturable absorber that shows self-pulsing. Along the dashed lines marked σ_s and σ_G the locking is prevented by keeping the frequency ratio at the "most irrational" values of the "silver mean" $\sqrt{2}-1$ and the "golden mean" $(\sqrt{5}-1)/2$. Following these lines, a direct transition from quasi-periodicity to chaos occurs. Because the intrinsic pulsing frequency shifts with modulation amplitude, paths of fixed ratio of pulsing frequency to modulation frequency are not straight lines. (From Winful et al., 1986.)

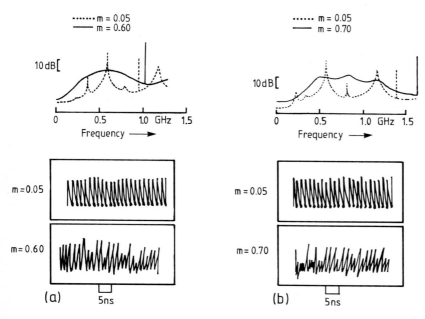

Fig. 6.2.10. Spectra of quasi-periodic pulsing (dashed lines) and chaotic pulsing (solid lines) and the corresponding measured time picture for (a) the "golden mean" and (b) the "silver mean"; m is the modulation index. (From Winful et al., 1986.)

In Fig. 6.2.10 the two cases are shown. The self-pulsing frequency was 0.6 GHz and the modulation frequency ca. 1 GHz and 1.5 GHz. Pulse shapes for the quasi-periodic and the chaotic range and their spectra are shown. The quasi-periodic spectrum is characterized by all linear combinations of the two frequencies, and the chaotic case by the appearance of strong broad-band spectra. The comparison with model calculations (Winful et al., 1986) yielded satisfactory agreement.

7 Lasers with Added Degrees of Freedom

Class B lasers, which are described by two equations (see Chapters 1 and 6), cannot display complex dynamics, such as chaotic time evolution (Chapter 3). However, simple ways exist to provide the additional degrees of freedom necessary for admitting such behavior even under single-mode conditions. In addition to the method described in the preceding chapter, with a saturable absorber inside the laser resonator, other methods make use of the time dependence of parameters of the laser. These are (i) temporal modulation of a laser parameter, such as resonator losses, resonator frequency or pump rate; (ii) addition of a feedback loop; and (iii) injection of an external optical field.

Class B lasers (such as CO_2 gas lasers and semiconductor or other solid-state lasers which are described by simple two-level rate equation models) with these external additions are the simplest laser systems showing low-dimensional chaotic behavior. All three cases are treated in this chapter.

7.1 Modulated Lasers

7.1.1 Theoretical Description

We describe CO_2 gas lasers and solid-state and NMR lasers. However, because they show many common features, we go into detail only in the first case.

a. CO$_2$ Gas Lasers

Consider a two-level homogeneously broadened single-mode gas laser in which the relaxation rate $\gamma_\perp$ of the polarization is much larger than the relaxation rate $\gamma_\parallel$ of the population inversion and the resonator damping rate κ. This approximately applies to a CO_2 laser, in which case $\gamma_\perp \approx 10^8 \text{ s}^{-1}$, $\gamma_\parallel \approx 10^3 - 10^4 \text{ s}^{-1}$ and $\kappa \approx 10^7 \text{ s}^{-1}$. This system may be described by means of the set of coupled equations (5.1.19) in which the induced dipole moment variable $p(t)$ can be adiabatically eliminated (Section 3.1.2.c). To perform this elimination, let us first omit from $E(t)$ and $p(t)$ the term oscillating at the optical frequency; assuming for simplicity resonant conditions, i.e. $\omega_c = \omega_{ab} = \omega$, these variables can be expressed in the form

$$E(t) = \tilde{E}(t)e^{-i\omega t}, \quad p(t) = \tilde{p}(t)e^{-i\omega t}. \tag{7.1.1}$$

By introducing eq. (7.1.1) into eqns. (5.1.19) and setting the time derivative of $p(t)$ equal to zero (adiabatic elimination), one obtains

$$p(t) = \frac{g}{\gamma_\perp}\tilde{E}(t)d(t). \tag{7.1.2}$$

If $p(t)$ is now replaced by eq. (7.1.2) in the first and third of eqns. (5.1.19) one obtains a pair of coupled equations:

$$\dot{\tilde{E}}(t) = \tilde{E}\left[-\kappa + \frac{Ng^2}{\gamma_\perp}d(t)\right] \tag{7.1.3}$$

$$\dot{d}(t) = \gamma_\parallel[d(t) - d_0] - \frac{4g^2}{\gamma_\perp}|\tilde{E}(t)|^2 d(t).$$

Since in the second equation $|E(t)|^2$ appears instead of $E(t)$, one can choose the intensity or the photon density instead of the field as the variable. To this end, and to obtain simple expressions, new dimensionless variables are defined:

$$I(t) = \frac{4g^2}{\gamma_\perp \gamma_\parallel}|\tilde{E}(t)|^2, \quad D(t) = \frac{d(t)}{d_0} \tag{7.1.4}$$

where I is proportional to the wave intensity and D to the total population inversion. With these variables eqns. (7.1.3) transform into the rate-equations set:

$$\dot{I}(t) = 2\kappa I(t)[-1 + A D(t)] \tag{7.1.5}$$
$$\dot{D}(t) = -\gamma_\parallel[D(t) - 1 + I(t)D(t)]$$

where

$$A = g^2 N d_0 / \kappa \gamma_\perp \tag{7.1.6}$$

is the pump rate or gain parameter.

Equations (7.1.3) or (7.1.5) describe class B lasers. They yield a non-trivial steady-state solution given by

$$\bar{I} = A - 1; \quad \bar{D} = 1/A. \tag{7.1.7}$$

A stability analysis (Section 3.2) of this solution reveals that above the first laser threshold ($A \geq 1$) it is always stable and that perturbations die away under damped oscillations (or "relaxation oscillations") of angular frequency (Statz et al., 1961; Tang, 1963; Haken and Sauermann, 1963; Tredicce et al., 1985):

$$\bar{\Omega} = [2\gamma_{\parallel}\kappa(A - 1)]^{\frac{1}{2}} \tag{7.1.8}$$

(frequency $\bar{f} = \bar{\Omega}/2\pi$); for a CO_2 laser, $\bar{\Omega} \approx 10^2\,\mathrm{kHz}$, see Arecchi et al. (1982) and Dangoisse et al. (1986).

To allow for more complex dynamics, a sinusoidal modulation of some parameters can be introduced, as was suggested in Yamada and Graham (1980), Scholz et al. (1981), Arecchi et al. (1982), Ivanov et al. (1982). We shall consider the specific cases of modulation of the resonator losses κ, the pump rate A and the resonator frequency ω_c.

Resonator loss modulation

In this case the parameter κ becomes time dependent in the following form:

$$\kappa \rightarrow \kappa(t) = \kappa_0(1 + m\cos\Omega t). \tag{7.1.9}$$

If we consequently redefine A as $A = g^2 N d_0/(\kappa_0 \cdot \gamma_{\perp})$, then eqs. (7.1.5) become

$$\dot{I}(t) = -2\kappa(t)I(t) + 2\kappa_0 A I(t)D(t) \tag{7.1.10}$$
$$\dot{D}(t) = -\gamma_{\parallel}[D(t) - 1] - \gamma_{\parallel}I(t)D(t).$$

This system of two equations is now non-autonomous, since one of the coefficients is time dependent (Section 3.1.1), and admits a dynamic behavior with a degree of complexity similar to that corresponding to a system with three degrees of freedom. This similarity becomes evident if one considers $\kappa(t)$ as a further variable and adds to the system (7.1.10) a third first-order equation:

$$\dot{\kappa}(t) = -\kappa_0 m\Omega \sin\Omega t. \tag{7.1.11}$$

In this way three degrees of freedom are evident.

By taking the derivative with respect to time of the first eqns. (7.1.10) and introducing into it $\dot{D}(t)$ from the second equation, one finds that this system is described by equations of the type of forced nonlinear oscillators (Tredicce et al., 1985).

Several workers have theoretically analyzed loss modulated gas lasers (Yamada and Graham, 1980; Scholz et al., 1981; Arecchi et al., 1982; Tredicce, Abraham et al., 1985; Tredicce, Arecchi et al., 1985; Midavaine et al., 1985; Dangoisse et al., 1987; Erneux et al., 1987; Glorieux, 1987; Mandel et al., 1988). Tredicce et al. (1985) studied the response of the steady-state solutions (7.1.7) to small perturbations by linearizing the system of eqns. (7.1.10). They found that small perturbations obey equations of the type encountered in the case of forced damped harmonic oscillators, $\bar{\Omega}$ being the resonant frequency (with κ replaced by κ_0 in eq. (7.1.8)) and $\gamma_\parallel A$ the damping rate. Thus an oscillation at the modulation frequency Ω occurs, which is linear in m and is resonantly enhanced when Ω approaches $\bar{\Omega}$. This means that a limit cycle in the phase space exists, whose stability properties can be analyzed by means of the Floquet theory (Section 3.2.1.b). Owing to the non-linear terms, the limit cycle turns out to be unstable for $\Omega \approx \bar{\Omega}$ even for small values of the modulation amplitude m, so that pulsations or other kinds of unstable behavior can appear. Indeed, by continuously varying the modulation frequency Ω in a zone close to $\bar{\Omega}$, Arecchi et al. (1982) theoretically predicted (and, as described below, experimentally observed) period-doubling sequences leading to chaos (Feigenbaum scenario; see Section 3.4.2.b), periodic windows within the cha-otic range and generalized bistability (see Section 3.3.3.c) between two co-existing attractors. Examples of these results are shown in Figs. 7.1.1 and 7.1.2, which corre-spond to calculations performed with parameter values appropriate to a CO_2 laser. Further, when the co-existing attractors both become chaotic, $1/f$-type power spectra may appear (Arecchi et al., 1982), as has been described in Section 3.3.3.c, in the presence of noise.

Modulation of the losses or of the pump rate also modulates the relaxation oscil-lation frequency $\bar{\Omega}$ (i.e. $\bar{\Omega}_k$ or $\bar{\Omega}_A$). This suggests interpreting the appearance of the above-mentioned unstable behavior for modulation frequencies close to $\bar{\Omega}$ as resulting from a resonant destabilization of the non-linear system by driving at the resonant frequency. Because of the non-linearities and mainly in the case of deep modulation, however, this interpretation might only represent a crude approximation to the actual dynamics.

Dangoisse and co-workers (Midavaine et al., 1985; Dangoisse et al., 1987) and Glorieux (1987) pointed out that a laser with parameters modulated at a frequency f

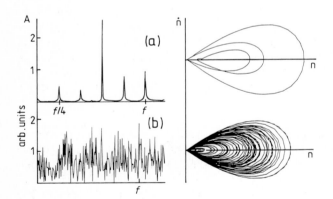

Fig. 7.1.1. Numerical calculation of CO_2 laser dynamics for the case of loss modulation. Right: phase-space projections in the $(n, \dot{n})$ plane (where n is the photon density (intensity)) showing (a) the $f/4$ subharmonic bifurcation ($f = \Omega/2\pi = 64.33$ kHz, $m = 0.02$) and (b) the strange attractor ($f = 78.8$ kHz, $m = 0.03$), both corresponding to a period-doubling cascade. Left: corresponding spectra (parameter values corresponding to a CO_2 laser: $\gamma_\parallel = 10^3$ s^{-1}, $\kappa_0 = 7 \cdot 10^7$ s^{-1}, $m = 2.0 \cdot 10^{-2}$, $A \approx 2$ ($Nd(t) = 2.0 \cdot 10^{11}$), $\gamma_\perp \approx 10^8$ s^{-1}. (From Arecchi et al., 1982.)

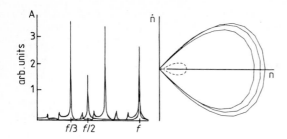

Fig. 7.1.2. As Fig. 7.1.1, for $f = 119.0$ kHz, $m = 2.0 \cdot 10^{-2}$. The phase-space portrait on the $(n, \dot{n})$ plane shows the existence of two independent attractors, which correspond to periodic orbits of frequencies $f/3$ (continuous line) and $f/2$ (dashed line). The corresponding spectra are super-imposed. This is an example of generalized bistability. In the experiment transitions between the two attractors caused by noise frequently occur. (From Arecchi et al., 1982.)

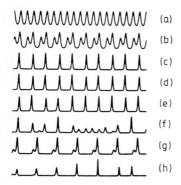

Fig. 7.1.3. Time dependence of the modulated laser intensity, calculated for $A = 1.1$, $\gamma_\parallel/2\kappa_0 = 2.083 \cdot 10^{-3}$ and $\Omega/2\pi = 400$ kHz. (a) $m = 0.01$, period 1; (b) $= m = 0.015$, period 2; (c) $m = 0.02$, period 4; (d) $m = 0.02127$, period 8; (e) $m = 0.022$, chaotic signal (quasi-periodicity 2T); (f) $m = 0.0246$, chaotic signal; (g) $m = 0.0337$, period 3; (h) $m = 0.0338$, chaos. (From Dangoisse et al., 1987.)

may respond not only at that frequency and its harmonics nf, but also at subharmonics f/n. They performed calculations showing the time dependence of the laser intensity (Fig. 7.1.3), which qualitatively reproduce their experimental findings. Very high degrees of modulation of the output signal can be observed, which are obtained with small modulation amplitudes m. They also traced bifurcation diagrams qualitatively similar to that of the logistic map (Section 2.2) and an attractor's map in the parameter space (m, A) (Dangoisse et al., 1987), and found the same dynamical features as Arecchi et al. (1982) (for instance, a period-doubling sequence as shown in Fig. 7.1.3); further, they observed crises (see Section 3.4.1) when a chaotic attractor touches an unstable periodic cycle, and they studied the influence of the rate of change of the driving parameters (Dangoisse et al., 1987; Glorieux, 1987).

More recently, Erneux et al. (1987) performed detailed analytical calculations also in the case of a small modulation amplitude, in which they found a rich variety of periodic solutions. In particular, they found large-amplitude periodic solutions oscillating at any of the harmonics or subharmonics of the modulation frequency Ω, and small-amplitude solutions oscillating at frequency Ω. Perturbations of these last oscillating states lead to slowly decaying quasi-periodic oscillations except when Ω coincides with $n\bar{\Omega}$ or $\bar{\Omega}/n$ ($n = 1,2,...$); for instance, when $\Omega \approx \bar{\Omega}$, where bistability of periodic solutions with the same frequency is observed, and $\Omega \approx 2\bar{\Omega}$, where subharmonic bifurcation may occur and the period of the oscillations suddenly doubles.

Erneux and co-workers (1987) also gave an intuitive explanation for the appearance of such a rich variety of dynamics. As already mentioned, in the absence of external modulation, the steady-state solution (7.1.7) is always stable above the laser threshold; actually, it is marginally stable in the limit $\gamma_{\parallel}/\gamma_{\perp} \rightarrow 0$ (or, in other words, it is a degenerate Hopf bifurcation point): the laser will relax towards its stable steady state through damped oscillations whose frequency and damping change as $(\gamma_{\parallel}/\gamma_{\perp})^{1/2}$ and $(\gamma_{\parallel}/\gamma_{\perp})$, respectively. Thus, a laser operating in this regime displays critical slowing down for any value of the pump parameter. We may therefore describe the resulting evolution as the small dissipative perturbation of a conservative system. This conservative system has bounded periodic solutions. As a result of the small damping rate, a small-amplitude external modulation may stabilize periodic solutions.

Mandel et al. (1988) also extended the previous analysis by Erneux et al. (1987) to take into account inhomogeneous broadening and detuning. Unlike in other kinds of lasers, they found the surprising result that these factors do not affect qualitatively the bifurcation diagrams. This can be explained in the following way. In most lasers, instabilities result from a Hopf bifurcation of the steady-state solution at the second laser threshold (see, for instance, Sections 5.2 and 5.3), where a characteristic exponent which plays the role of the relaxation rate for the damped oscillations (see Section 3.2.1.a) goes to zero when a control parameter is continuously varied. As the dependence of the relaxation rate on the control parameter is in general strongly affected by inhomogeneous broadening and detuning, the unstable behavior also becomes affected by these factors. On the other hand, with loss-modulated lasers the instability of the steady state results from a resonance effect in which weakly damped oscillations are excited by the external modulation. Hence a commensurate ratio is required between the driving frequency and the damped resonance frequency independent of additional effects such as inhomogeneous broadening or detuning.

Dangoisse et al. (1987) and Glorieux (1987) studied numerically the influence of a non-zero sweep rate of the modulation amplitude m on the bifurcation diagrams.

Pump modulation

In this case the modulated parameter is the pump rate or gain parameter A appearing in eqns. (7.1.5), which becomes

$$A \rightarrow A(t) = A_0(1 + l\cos\Omega t) \qquad (7.1.12)$$

with A_0 given by eq. (7.1.6).

As in the previous case, the system (7.1.5) becomes non-autonomous. Different periodic and chaotic solutions can again appear. Tredicce et al. (1985) performed in this case a study similar to that mentioned above for the case of loss modulation. They found that small perturbations of the steady-state solution behave very similarly in both cases. $\bar{\Omega}$ (defined as in eq. (7.1.8) with A_0 instead of A) is again the resonant frequency and $\gamma_\parallel A_0$ the damping rate. The main difference lies in the dependence on the respective modulation depths m and l. The ratio m/l that produces the same effects for the two cases is

$$m/l \approx [A_0\gamma_\parallel/(A_0 - 1)2\kappa_0]^{1/2} \qquad (7.1.13)$$

when $\Omega \approx \bar{\Omega}$. For CO_2 or solid-state lasers with $A_0 \approx 2$, this ratio is ca. 10^{-2}, i.e. the degree of modulation of the pump must be much larger than the degree of modulation of the resonator losses. The presence of non-linearities, however, introduces differences. Intuitively, one would indeed expect differences, since e.g. cutting the pump abruptly would obviously result in relaxation oscillations, while increasing the resonator losses to a high value would simply cut the laser emission to zero.

A numerical study of pump-modulated lasers was performed by Lauterborn and Steinhoff (1988), who classified the bifurcation structure of the system through the concepts of non-linear resonances and torsion numbers. They showed the close relationship between the pump-modulated laser rate equations and the Toda oscillator.

Arecchi et al. (1988) found a difference between pump and loss modulation when a linear sweep of the modulation amplitude was performed. They observed only in the latter case a "delayed bifurcation", (i.e. a dynamic stabilization) of the zero-intensity solution, which remains stable for a pump parameter domain larger than with an adiabatic increase of the pump parameter (Mandel and Erneux, 1984). They explained theoretically this observed behavior in a CO_2 laser by considering the four-level molecular model which takes into account the coupling between the two resonant levels and the respective rotational manifolds (see Chapter 6).

Finally, pump modulation of an N_2O laser containing ammonia as a saturable absorber was recently investigated by Hennequin et al. (personal communication) (Chapter 6).

Resonator frequency modulation

When the length of the optical resonator is modulated, its frequency ω_c is also changing at the modulation frequency. An analysis similar to that of eqns. (7.1.1)–(7.1.6) using the two-level model for the CO_2 molecular medium (Chapter 6) and allowing for any value of the resonator detuning $\delta = (\omega_c - \omega_{ab})/\gamma_\perp$ leads to the result (Midavaine et al., 1985; Dangoisse et al., 1987; Glorieux, 1987)

$$\dot{I}(t) = 2\kappa \cdot I(t)[-1 + A D(t)/(1 + \delta^2)] \tag{7.1.14}$$
$$\dot{D}(t) = \gamma_{\parallel}[D(t) - 1 + I(t)D(t)/(1 + \delta^2)].$$

The modulation appears in the time dependence of δ:

$$\delta \to \delta(t) = \delta_0(1 + m\cos\Omega t) \tag{7.1.15}$$

and, as in the previous cases, the system (7.1.14) is non-autonomous.

Dangoisse et al. (1987) performed numerical calculations and found that the bifurcation diagrams do not show any qualitative difference from those obtained with loss modulation, except for the higher modulation index needed in the frequency modulation case (about one order of magnitude with their set of parameters). The degree of frequency modulation necessary to destabilize the laser becomes progressively smaller compared with the loss modulation, with increasing detuning.

b. Semiconductor Diode Lasers and Solid-State Lasers

As already described (Ch. 1 and Sect. 6.1.2), in solid-state lasers the polarization can be adiabatically eliminated and a rate-equation model taking as variables the injection carrier (n_e) and photon (n_p) densities can be used. Consider the standard simplified rate equations for a semiconductor laser:

$$\frac{dn_p}{dt} = -\frac{n_p}{\tau_p} + A(n_e - n_e^0)n_p + \beta\frac{n_e}{\tau_e}$$
$$\frac{dn_e}{dt} = P - \frac{n_e}{\tau_e} - A(n_e - n_e^0)n_p \tag{7.1.16}$$

where τ_e and τ_p are the carrier and photon lifetimes, respectively, A represents the gain parameter, β is the spontaneous emission factor, $P = I/eV$ (I is the injection current, e is the electron charge and V is the active volume) and n_e^0 is the minimum carrier density required for positive gain. An example of extended equations taking into account further experimental complications was given in eqs. (6.1.5).

The structure of eqns. (7.1.16) is very similar to that of eqns. (7.1.5) describing a CO_2 gas laser, except for the term $\beta n_e/\tau_e$, which does not exist (as it is negligible) in the gas laser. Hence the influence of external modulation can give rise to common features in both cases.

In the case of semiconductor lasers, the study of the influence of modulation of the injected current is of interest for applications in optical communications:

$$I \to I(t) = I_0(1 + m\cos\Omega t).$$

Here not only small modulation depths ($m \leqslant 1$), but also large ones ($m \lesssim 1$) are used. The first prediction of a period-doubling route to chaos in a semiconductor laser when m increases was made by Lee et al. (1985). Hori et al. (1988) rewrote eqns. (7.1.16) in the form of a second-order differential equation for a driven non-linear oscillator (in a similar way to Tredicce et al. (1985) and Arecchi and co-workers (1984) for a CO_2 laser), with terms representing the damping, restoring and driving forces. They investigated the origin of chaos and found that it lies in the non-linearity of the restoring force. They also found that an increase in the spontaneous emission factor β affects the non-linearities and reduces the possibilities of the appearance of chaos. It is for this reason that chaos is more readily observed in CO_2 gas lasers (where β is negligible) than in semiconductor lasers with a large β.

Chen et al. (1985) took into account the influence of the intrinsic laser amplitude fluctuations brought about by the quantum nature of spontaneous emission. They modeled these fluctuations by adding to eqns. (7.1.16) Langevin noise terms $F_p(t)$ and $F_e(t)$, which are delta-correlated Gaussian random variables with zero mean. They found that the inclusion of quantum noise leads to large amplitude fluctuations in the modulated output, which inhibit the development of the higher subharmonics in the predicted period-doubling sequence, so that in their case only periods T, $2T$ and sometimes $4T$ are observed (see Section 3.4.2). Further, deterministic chaos appears mixed with these intrinsic fluctuations.

Shore (1988) emphasized the amplification properties of modulated semiconductor lasers (i.e. amplifications of optical signals at side-bands of the optical laser frequency) with the aim of finding practical applications for non-linear dynamical phenomena. His approach is based on the work of Wiesenfeld and McNamara (1986), who analyzed the amplifying properties of a system near a bifurcation point. They found that in the case of a Hopf bifurcation the response curve of the system to the modulation of a control parameter is given by the function

$$H(\Omega) = [(\Omega - \omega_H)^2 + \varepsilon^2]^{-1} \qquad (7.1.17)$$

where $\lambda_\pm = \varepsilon \pm i\omega_H$ are the characteristic exponents associated with the Hopf bifurcation (Section 3.4.1.a). As in this kind of bifurcation the real part ε crosses zero, large amplification could in principle be obtained. Shore (1988) studied this possibility in the case of the Hopf bifurcation associated with a modal switching instability in a gain-guided stripe-geometry laser diode.

Other kinds of solid-state lasers have also been investigated theoretically. Otsuka (1978) and Kubodera and Otsuka (1981) considered the case of a two-mode laser-diode pumped $LiNdP_4O_{12}$ laser with deep pump modulation, in which a spatial dependence of the population inversion density brought about by the spatial hole burning was partially taken into account. Khandokhin and co-workers (Polushkin et al., 1983; Khandokhin et al., 1984) studied a single-frequency YAG:Nd^{3+} traveling-wave ring laser subjected to loss modulation, in which the existence of a weak counter-propagating wave was taken into account (see also Ivanov et al., 1982).

Current modulation in semiconductor lasers with saturable absorbers has been analyzed by Kawaguchi (1984) using eqns. (6.1.5). A period-doubling route to chaos was found for deep modulation when the modulation frequency was continuously increased.

7.1.2   Experiments

The dynamics of lasers with modulated parameters has been studied experimentally on a number of laser systems. CO_2 lasers have been studied with modulated gain, modulated resonator frequency and modulated resonator losses. Continuous Nd^{3+} solid-state lasers excited by lasers and semiconductor diode lasers have been investigated by modulating their gain. These lasers all belong to "class B" and are therefore described by the same equations.

The ruby NMR "laser" operates on a spin-flip transition in the radio-frequency range and is therefore sometimes termed a RASER. Although it is not a laser in the sense of a light generator, it is described by three equations identical with (5.1.19) (possibly showing some inhomogeneous broadening). This system would be suitable for observing the Lorenz/Haken instability. Since, however, a high enough excitation has so far not been reached, experiments and theoretical modeling have been performed for the case of modulated parameters, or an injected external field (see Section 7.3). If a variable is to be adiabatically eliminated for the NMR laser then it is the field, because the resonator damping time is very short compared with polarization and population damping times. Therefore, as for the class B lasers, for the NMR laser the number of equations is two. However, for the class B laser it is the field and population equations that remain, whereas for the NMR laser it is the polarization and population equations. Consequently, the phenomenologies of class B and NMR lasers could be different. Class B lasers can be pictured as non-linear damped oscillators, when periodically driven, known to exhibit chaotic dynamics.

CO_2 Lasers, Loss Modulation

Most notably this system, as expected of a driven non-linear oscillator, shows a period-doubling transition to chaos (Tredicce et al., 1986). Figure 7.1.4 shows spectra and phase space projections for illustration. The loss of the laser is modulated by an electro-optic modulator at a frequency near the relaxation oscillation frequency itself or near its subharmonics. The modulation amplitude necessary to destabilize the laser output is lowest near the relaxation frequency itself.

Periodic "windows" within the chaotic range up to period-10 could be observed. The period-doubling transition to chaos at first appears to indicate typical dynamics of the one-dimensional logistic map; however inspection of the chaotic range reveals otherwise. Figure 7.1.5 shows bifurcation diagrams measured by stroboscopically viewing the laser output. The laser intensity is sampled at a fixed point in every period of the modulation. The samples are recorded as a function of the loss modulation amplitude.

It is apparent that the bifurcation diagram is different for increasing and decreasing modulation amplitude. The reason for this hysteresis is multiple coexisting states. As mentioned, the laser can be destabilized by interacting with all subharmonics of the relaxation oscillation. With every subharmonic is associated a complete bifurcation diagram as in Fig. 2.2.1 with their instability points; of course, at different modulation amplitude values, and all these coexist.

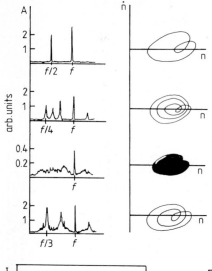

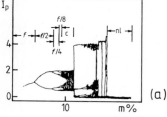 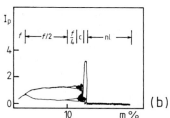

Fig. 7.1.4. Intensity spectra of a CO_2 laser whose loss is modulated at a frequency near the CO_2 laser relaxation resonance frequency. Periods two and four are visible, followed by a chaotic spectrum and the well known period-three window. Also shown is a "phase-space projection": the dependence of $\dot{I}$ on I (I = laser intensity) (From Arecchi et al., 1982.)

Fig. 7.1.5. Bifurcation diagrams of a loss-modulated CO_2 laser showing the coexistence of attractors by first increasing (a) and then decreasing (b) the modulation amplitude. Coexistence of periodic with chaotic states is seen. Hysteresis of non-lasing state is also evident. (From Tredicce et al., 1986.)

Coexistence of multiple states (multistability) leads to hysteresis, as is usual in the field of bistability − the simplest case. Of course, in the present case chaotic states coexist with chaotic or periodic states, leading to various kinds of "crises" (Chapter 3), i.e. sudden expansions, contractions or disappearances of solutions when two states coalesce.

The phenomenon of "crisis" has been studied in more detail with a high-pressure CO_2 laser (Dangoisse et al., 1986). Figure 7.1.6 shows the sudden expansion of the chaotic attractor within the reverse, part-chaotic, period-doubling cascade of the frequency-modulated laser.

In Fig. 7.1.7, the chaotic attractor coalesces with a period-three periodic state. The result is a sudden expansion of the chaotic attractor. It is noted that remains of the period-three state continue to exist, as seen in the density of points of the expanded attractor.

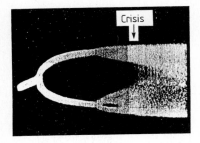

Fig. 7.1.6. Bifurcation diagram of a loss modulated CO_2 laser showing period doublings, chaos and a crisis corresponding to an expansion of the attractor which fills the space between the bands of the part chaotic reverse ("period dedoubling") cascade. (From Dangoisse et al., 1986.)

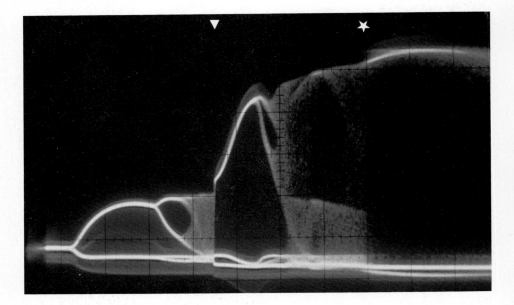

Fig. 7.1.7. Bifurcation diagram of a loss modulation CO_2 laser showing a boundary crisis (∇) and interior crisis (*). (From Dangoisse et al., 1986.)

A certain choice of the states that coexist can be made by varying the laser resonator tuning with respect to the laser gain line. For a class B laser this changes primarily the pump parameter. The bifurcation diagrams in Fig. 7.1.8 are seen to change drastically with laser tuning.

Reaching the asymptotic solution (state of the laser) requires a finite and sometimes substantial time. Since parameters have to be swept in order to record bifurcation diagrams, the laser may not be able to "adiabatically" follow the parameter change, i.e. to pass through a series of asymptotic solutions. In this case an upsweep and a downsweep of some parameter produces different bifurcation diagrams which, apart from the coexistence of asymptotic solutions with its related hysteresis effects, can produce different laser behavior for the same parameter value ("dynamically induced

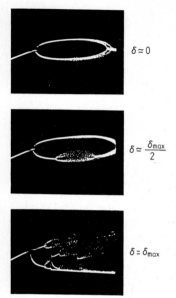

$\delta \simeq 0$

$\delta \simeq \dfrac{\delta_{max}}{2}$

$\delta = \delta_{max}$

Fig. 7.1.8. Change of bifurcation diagram of a modulated CO_2 laser with detuning. $\delta = 0$, resonant tuning; $\delta = \delta_{max}$, 200 MHz detuning from line center. (From Midavaine et al., 1985.)

bistability"). Figure 7.1.9 shows three bifurcation diagrams for up- and downsweeping of the modulation amplitude for different sweep frequencies. The dependence of the observed diagram on the sweep rate is clearly observable.

Figure 7.1.10 gives a picture of the exponential divergence of time evolutions for near initial conditions. The curves were chosen to coincide at zero time within the measurement uncertainty. They are seen to separate with time. From such measurements, entropies or Lyapunov exponents might be obtained.

CO_2 lasers, gain modulation

An experiment on gain modulation of a CO_2 laser (Biswas et al., 1987) showed different phenomena. The situation is different from the aforementioned cases since the modulation frequency in this experiment is two orders of magnitude lower than the relaxation resonance frequency. Figure 7.1.11 shows the regions of different dynamics in the detuning of the laser resonator with respect to the gain line center. Figure 7.1.12 shows an intermittent transition to chaos in the time picture. The results by Biswas et al. (1987) indicated that chaotic laser dynamics may be already caused by slight technical imperfections of a laser as the pump modulation in this case was brought about by an insufficiently filtered current supply for the laser discharge.

The 100 Hz modulation of the laser current (remains of the mains AC voltage) at which chaotic laser dynamics occur is only 1%, a value usually considered "good" for CO_2 laser power supplies.

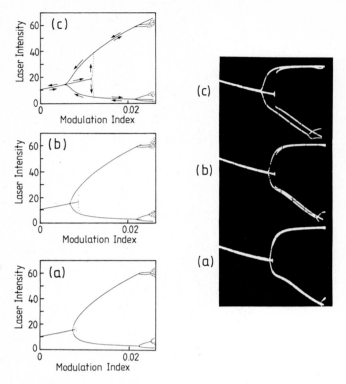

Fig. 7.1.9. Theoretical (left) and experimental (right) bifurcation diagrams of a modulated CO_2 laser for different sweep speeds (a, b, c). (From Dangoisse et al., 1987.)

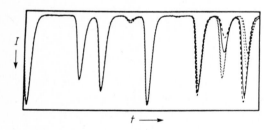

Fig. 7.1.10. Divergence of trajectories of a chaotic, modulated CO_2 laser. The three traces of irregular pulsing start at almost the same initial conditions. The slow divergence with time shows a small Lyapunov exponent. (From Dangoisse et al., 1987.)

Solid-state lasers, gain modulation

Optically pumped solid-state lasers have been found to exhibit chaotic dynamics when the pump strength is modulated.

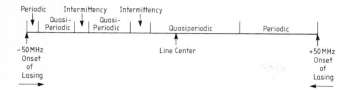

Fig. 7.1.11. The different ranges of the (gain modulated) CO_2 laser tuning in which the laser shows different dynamics. (From Biswas et al., 1987.)

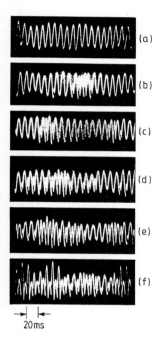

20ms

Fig. 7.1.12. Onset of chaos through intermittency. The regular pulsing is interrupted by more and more irregular bursts when the laser resonator is progressively detuned (a–f). (From Biswas et al., 1987.)

The first indication was given by Kubodera and Otsuka (1981). The laser used an $LiNdP_4O_{12}$ crystal as the active medium, excited by light from a semiconductor laser at 800 nm. The aim was to generate continuous trains of relatively high-power laser light pulses by modulating the pump strength. Together with these pulses, pulses at the first subharmonic of the modulation frequency ($P2$) were observed. Pulses at one third of the modulation frequency and "unstable multispike pulses" were mentioned, the latter probably indicating chaotic dynamics.

With the aim of obtaining clearer proof of chaotic laser dynamics, a similar material (NdP_5O_{14}) was used, pumped by a modulated Ar^+ laser (Klische et al., 1984). A clear period-doubling transition (Fig. 7.1.13) followed by broad-band spectra unambiguously demonstrated chaos, including the well known periodic $P3$ and $P5$ windows in the chaotic range, when the Ar^+ pump laser power was modulated at a frequency approximately corresponding to the relaxation resonance.

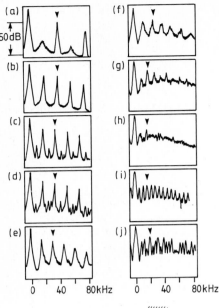

Fig. 7.1.13. Intensity spectra of an NdP_5O_{14} solid-state laser modulated near twice the relaxation oscillation resonance frequency of the laser. The solid-state laser is pumped by an Ar^+ laser whose intensity is periodically changed by an electro-optic modulator. As the modulation frequency is moved towards the relaxation resonance, pulsing and period doubling ending in chaotic dynamics ((h), fully developed chaos) is observed. Within the chaotic region the well known $P3$ and $P5$ "windows" are observed ((i) and (j), respectively). The modulation frequency is marked. (From Klische et al., 1984.)

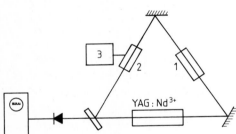

Fig. 7.1.14. Loss modulated Nd:YAG ring laser. 1, Optical isolator for uni-directional transmission; 2, electro-optic modulator; 3, modulator driver. (From Khandokhin and Khanin, 1984.)

Observations on a fiber laser with modulated pumps showed similar effects (Phillips et al., 1987). The laser used a silica monomode optical fiber doped with Nd^{3+}. The pump source was again a 800 nm semiconductor diode laser. No clear period-doubling sequence was reported, but $P2$ and $P3$ subharmonic pulsing close to a region of completely irregular pulses was mentioned as an indication of the occurrence of chaos.

Chaotic dynamics were also observed in an Nd:YAG unidirectional ring laser (Fig. 7.1.14) (Khandokhin et al., 1984). The sequence of pulse shapes calculated with a realistic model is shown in Fig. 7.1.15 for increasing modulation depth. The laser output power follows the modulation linearly for small modulation depth. For stronger modulation the shape of each pulse develops two maxima. The laser then switches to sharp, single pulses, period-doubles and shows chaotic emission. After the chaotic range, sharp single pulses at half the modulation frequency appear. Suffice it to say that the obser-

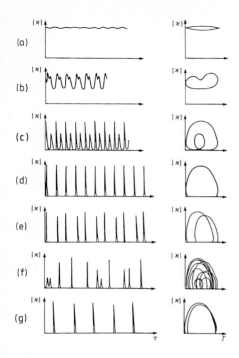

Fig. 7.1.15. Numerical results of a model of a modulated Nd:YAG ring laser. The left-hand side shows the laser intensity I as a function of time and the right-hand side the phase-space projection in the $I - \dot{I}$ plane. The modulation frequency is 0.4 times the relaxation resonance frequency. From a to g the modulation of the resonator losses is increased from 0.003 to 0.06. Period doubling and chaotic emission are evident. (From Khandokhin and Khanin, 1984.)

vations almost quantitatively reproduce the theoretically calculated pulses. Since no simple period-doubling transition (with the period-three windows) is found, one would again presume that this system shows multistability (coexisting states), as observed with CO_2 lasers.

Semiconductor lasers

The above findings suggested that similar effects might occur in semiconductor diode lasers. The occurrence of laser instabilities of the kind observed with Nd lasers could cause problems in optical communication where modulated lasers serve as transmitters. In this application, modulation frequencies comparable to the laser relaxation oscillation frequency are used.

Subharmonic pulsing of modulated diode lasers is in fact observed with multimode diode lasers (Grothe et al., 1976) and lasers emitting multiple modes under modulation (Weiss and Cho, unpublished). Figure 7.1.16 shows period doubling, tripling and quadrupling for the latter case.

Since single-mode lasers (distributed feedback lasers) will probably be used in communication systems, observations have been made on them (Ohtsu et al., to be publ.). Figure 7.1.17 shows a spectrum of such a modulated laser. The modulation frequency is twice as high as the laser relaxation oscillation frequency. A noisy period-two oscillation is observed but no higher subharmonics. This result was confirmed using different types of diode lasers (Chen et al., 1985). As mentioned above (Section 7.1.1),

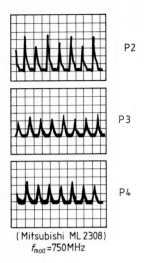

P2

P3

P4

(Mitsubishi ML 2308)
f_{mod} =750MHz

Fig. 7.1.16. Output power as function of time for a current-modulated diode laser. Second, third and fourth subharmonic pulsing can be seen. (From Weiss and Cho, unpublished work.)

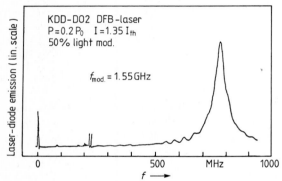

KDD-D02 DFB-laser
$P=0.2P_0$ $I=1.35I_{th}$
50% light mod.

$f_{mod.}=1.55$ GHz

Laser-diode emission (lin. scale)

0   500   MHz   1000
$f \longrightarrow$

Fig. 7.1.17. Intensity spectrum of a 1.55 μm DFB diode laser current-modulated at 1.55 GHz. The strong line at 775 MHz shows the onset of subharmonic pulsing. (From Ohtsu et al., to be published.)

models of diode lasers do show the complete period-doubling sequence ending in chaotic dynamics (Lee et al., 1985; Hori et al., 1988). The large spontaneous emission, however, typical of short-wavelength semiconductor lasers, causes large fluctuations which destroy the higher subharmonics (Section 7.1.1). The period-doubling cascade of modulated diode lasers is therefore prematurely truncated by the noise (Schuster, 1988). In fact, the noisy period-two emission when viewed in the time picture shows the characteristics of intermittent chaotic dynamics (Chen et al., 1985), brought about at the early stage of the $P2$ bifurcation by the spontaneous emission.

NMR laser

As mentioned, the NMR ruby laser ("RASER") shows chaotic dynamics when parameters are modulated (Brun et al., 1985). The NMR laser uses the magnetic dipole moments of the aluminum nuclei of ruby. The ruby crystal is mounted in the coil of an LC circuit which acts as the laser resonator and a static magnetic field B_0 is applied. The energy levels are $E_m = -mghB_0$, between which $\Delta m = \pm 1$ magnetic transitions are possible. A population inversion between two adjacent Zeeman levels of 10% is necessary to reach the laser threshold. The thermal population difference is reduced by cooling the crystal (and the coil) to liquid helium temperature. Population inversion is then achieved by exciting an electron-spin resonance transition with a microwave field. By spin-spin interaction the polarization of the electron spins is then transferred to the nuclear spins. The 10% population inversion corresponds to a negative temperature of some milli-Kelvin. The corresponding population inversion then leads to an undamped oscillation of the LC circuit. The output of the laser can be measured as the voltage $V(t)$ across the LC circuit.

Single-mode oscillation is possible since the nuclear Zeeman levels are shifted by electric quadrupole interaction with the crystal field. Five transition frequencies result to which the LC circuit can be tuned. If the inversion is not too high, multi-frequency emission can be prevented.

The tuned NMR laser is described by the laser equations (5.1.19). A typical set of damping constants is $\kappa = 4 \cdot 10^5 \text{ s}^{-1}$, $\gamma_\perp = 3 \cdot 10^4 \text{ s}^{-1}$, $\gamma_\parallel = 10 \text{ s}^{-1}$. The pump parameter r lies between 1 and 1.5.

The NMR laser is thus a "bad-cavity" system: $\kappa \gg \gamma_\perp + \gamma_\parallel$ (in fact, a "very bad-cavity" system), and thus in principle suitable for observing the Lorenz instability if the necessary pump strength is reached.

As the field damping rate is large compared with the other damping rates, the laser field may be adiabatically eliminated (other than for a class B laser, where the polarization is eliminable), reducing the system phase space dimension to two.

Various quantities can be modulated to increase the phase space dimension: LC circuit loss, the static magnetic field, or the spin inversion. In addition, the NMR linewidth can be modulated by a weak time-dependent gradient of the magnetic field. This is called T_2-($\gamma_\perp$-) modulation and was used in the experiments.

The time-dependent NMR laser outputs for increasing modulation amplitude was Fourier-analysed. The spectra show clearly that a period-doubling transition to chaos occurs. Figure 7.1.18 shows the $P8$ subharmonic and chaotic oscillation of the NMR laser output power in the time-picture and in a phase-space projection.

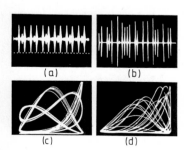

(a) (b)

(c) (d)

Fig. 7.1.18. Output power of the "linewidth-modulated" NMR laser for two different degrees of modulation. (a, b) Time pictures; (c, d) phase-space projections; (a, c) $P8$ pulsing; (b, d) chaotic pulsing. (From Brun et al., 1985.)

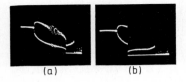

Fig. 7.1.19. Bifurcation diagram measured with a stroboscopic technique on the "linewidth-modulated" NMR laser. $f = 72$ Hz; (a) up and (b) down scan of modulation amplitude. The difference in the two pictures shows dynamical states coexisting. (From Brun et al., 1985.)

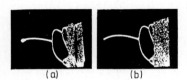

Fig. 7.1.20. As Fig. 7.1.19, except modulation frequency $= 70$ Hz. Coexisting dynamical states; (a): up scan, (b): down scan. (From Brun et al., 1985.)

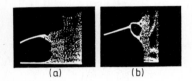

Fig. 7.1.21. Difference in bifurcation diagrams for $f =$ (a) 108 Hz and (b) 66 Hz (up scans). (From Brun et al., 1985.)

Multistability (coexisting states) is found experimentally. Figure 7.1.19 shows the measured bifurcation diagram for increasing and decreasing modulation amplitude. The fact that the two diagrams are different clearly indicates multistability.

A slightly slower rate of change of modulation amplitude (Fig. 7.1.20) yields, in contrast to Fig. 7.1.19, a fully chaotic region. This shows that the change of parameter for these conditions is too fast for the system to pass through equilibrium states, and in addition it shows the coexistence of different chaotic attractors. Figure 7.1.21 shows a bifurcation diagram at a slower rate of change of modulation amplitude (a) and at a faster rate (b). The difference is again clearly visible. Sudden expansions of the attractor are encountered, giving evidence of "crises" (Chapter 3). The experimental results (a) and (b) are qualitatively reproduced in the numerical solutions (Fig. 7.1.22).

Intermittent pulsing as onset of chaos has been observed in the "period-five" window (Fig. 7.1.23). The change in laser dynamics as the modulation frequency is changed at constant amplitude is shown in Fig. 7.1.24, where the maxima of the laser power of each modulation period are plotted against the modulation frequency (normalized to the relaxation oscillation frequency). The jumps are typical for non-linear driven os-

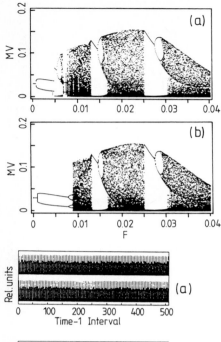

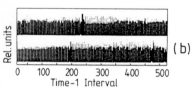

Fig. 7.1.22. Calculated bifurcation diagrams for the "linewidth-modulated" NMR laser for swept modulation amplitude: (a) slow sweep; (b) six times higher sweep speed. The "windows" created as a consequence of "crises" are apparent. (From Brun et al., 1985.) MV: laser output, F: modulation amplitude.

Fig. 7.1.23. Time picture of NMR laser pulsing with "line-width modulation" showing (a) regular pulsing and (b) intermittency. (From Brun et al., 1985.)

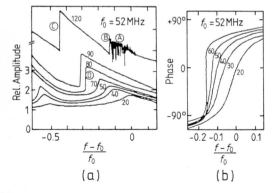

Fig. 7.1.24. Experimental NMR laser response to linewidth modulation at constant modulation amplitude and varying modulation frequency. (a) Pulse peaks and (b) the shift of the laser pulses with respect to the modulation period T. (From Brun et al., 1985.)

cillators and show frequency bistability. Chaotic dynamics occurs in the range marked by A in Fig. 7.1.24a. The pulsing at the boundary between regular and chaotic pulsing, B (Fig. 7.1.24a), indicates that the transition to chaos is intermittent. The transitions at C and D, where the laser jumps from one stable state to the other, show that at the transition points the pulsing shapes change. The hysteresis expected for driven non-linear oscillators which show multistability and jumps between the stable states was also confirmed. Figure 7.1.25 gives a magnification of the output pulses at points A and B of Fig. 7.1.24. Figure 7.1.26 shows an upward and downward sweep of the modulation frequency corresponding to the conditions of Fig. 7.1.24 (modulation period $T = 80$ μs).

7.2 Feedback Lasers

Negative electrical feedback is frequently used in laser design. For instance, output power stability in gas lasers is improved by controlling the pump strength with an electrical signal provided by the detected output intensity. Optical feedback is also used for noise reduction in semiconductor lasers, and the laser frequency can be actively stabilized by feedback control.

From the point of view of laser dynamics, introduction of feedback provides a further degree of freedom which may increase considerably the variety of dynamics that can be expected in class B lasers. We describe this problem by focusing attention on the case that has been studied in detail, i.e. a CO_2 laser with electrical feedback controlling resonator losses. The case of controlling resonator length and the case of optical feedback in semiconductor lasers are discussed.

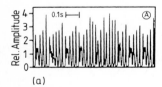

(a)

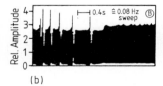

(b)

Fig. 7.1.25a, b. Magnification of the output pulses of NMR laser at points A and B of Fig. 7.1.24. (From Brun et al., 1985.)

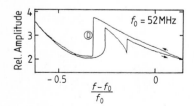

Fig. 7.1.26. Hysteresis for up and down sweeps of the modulation frequency corresponding to point D of Fig. 7.1.24 ($T = 80$ μs).

7.2.1 Theoretical Description

Let us consider a single-mode class B laser, which in a simple two-level model with resonant tuning is described by eqns. (7.1.5). If, according to the schematic diagram of the experiment shown in Fig. 7.2.1, an electric feedback loop is used which controls the resonator losses by a signal dependent on the laser intensity, a new degree of freedom appears. The control voltage $V(t)$ represents a new independent variable of the system. The resulting set of coupled equations is

$$\dot{I}(t) = 2\kappa_0 I(t)[-1 - \alpha \sin^2 V(t) + A D(t)]$$
$$\dot{D}(t) = \gamma_{\parallel}[D(t) - 1 + I(t)D(t)] \qquad (7.2.1)$$
$$\dot{V}(t) = -\beta[V(t) - B + r I(t)]$$

(Arecchi et al., 1986) where $2\kappa(V) = 2\kappa_0(1 + \alpha \sin^2 V)$ represents the resonator losses (for the field intensity) controlled by the voltage V, κ_0 is the resonator loss without the additional intensity-dependent loss and α is a constant describing the loss modulator. The $\sin^2$ dependence is due to the electro-optic modulator used (see Fig. 7.2.1), β is the damping constant of the feedback loop, B is an additional voltage added to the modulation voltage and r is the ratio between the intensity detected and the voltage applied to the loss modulator. The voltage V is usually given in angular units, that is, it is normalized to V_0/π, where V_0 is the $\lambda/2$ modulator voltage. The degree of feedback coupling is controlled by the parameter r.

Because the system described by eqns. (7.2.1) has three degrees of freedom, a rich dynamic behavior, including chaotic time evolution, is possible. This occurs when the time constant of the feedback loop (controlled by β) is of the same order as that of the other two relevant variables (when the feedback loop is so fast that it provides an almost instantaneously adapted loss coefficient, it does not modify the phase-space topology).

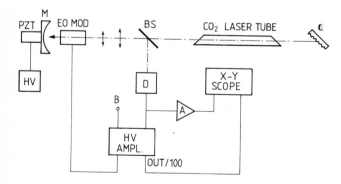

Fig. 7.2.1. Experimental set-up. M, total reflecting mirror mounted on a piezoelectric (P.Z.T). drive; E.O.MOD, electro-optic modulator; B.S., ZnSe beam splitter; G, grating; D, HgCdTe detector; B, bias voltage; H.V., high-voltage amplifier; A, amplifier. (From Arecchi et al., 1987.)

Equations (7.2.1) yield up to three coexisting stationary solutions whose stability properties determine the dynamical features of the system. These stationary solutions $(\bar{I}, \bar{D}, \bar{V})$ satisfy the condition

$$B = r\bar{I} + \arcsin\left[\frac{A/\alpha}{1+\bar{I}} - \frac{1}{\alpha}\right]^{1/2} \tag{7.2.2}$$

which is represented in Fig. 7.2.2 for fixed values of A and α and several values of r. It is seen that for a wide range of r values three stationary solutions (labeled 0, 1 and 2) coexist.

A linear stability analysis (Section 3.2.2.a) of these stationary solutions reveals the following properties (Arecchi, Gadomski et al., 1988; Arecchi, Lapucci et al., 1988; Arecchi et al., 1987):

(i) The solution 0 is the zero-intensity solution, and the corresponding fixed point in the three-dimensional phase space is a saddle node. It has three real eigenvalues, two negative and one positive, which means two stable directions (or, in other words, a two-dimensional stable manifold; see Section 3.2.1.a) and one unstable direction (or a one-dimensional unstable manifold), as in the case of the Lorenz model (Section 2.1).

(ii) The solution 1 corresponds to the steady-state output, and the corresponding fixed point in the phase space changes from a stable point to a limit cycle through a Hopf bifurcation (Section 3.4.1.a).

(iii) The fixed point corresponding to solution 2 is a saddle focus (Section 3.2.1.a), i.e. it has a real positive eigenvalue (unstable direction) and two complex conjugate eigenvalues with a negative real part (stable two-dimensional focus; see Sections 3.2.1.a and 3.2.2.a). Because the real positive exponent is larger than the magnitude of the negative real part of the complex ones, the conditions for the appearance of Shilnikov chaos are met (see Sections 3.3.3.b and 3.4.1).

Indeed, as is described below, the first experimental observation of this kind of

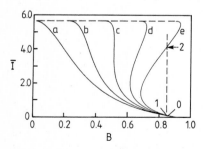

Fig. 7.2.2. Laser with feedback. Stationary intensity $\bar{I}$ versus the bias voltage B for a CO_2 laser with $A = 6.667$ and $\alpha = 9.0$. Curves a, b, c, d and e refer to $r = 0.0, 0.04, 0.08, 0.12$ and 0.16, respectively. The horizontal dashed line corresponds to stationary solutions with $V = 0$. Points 0, 1 and 2, indicated by arrows, are the stationary points for $B = 0.838$ and $r = 0.16$. (From Arecchi, Gadomski et al., 1988.)

dynamics in laser physics was made by Arecchi et al. (1987) on a CO_2 laser with electrical feedback.

Over a wide range of parameter values the dynamic behavior is characterized by "global" features (see Section 3.4.1) in the phase space, related to the presence of the three coexisting unstable fixed points. This is shown in Fig. 7.2.3, where a closed orbit visiting successively the neighborhoods of the three points characterizes the evolution of the system with time. Hence, different regimes can be observed when a control parameter is varied and the relative attraction of the different fixed points changes. Competition among different instabilities, displaying transitions from limit cycles generated in the Hopf bifurcation to local chaos (generated by a period-doubling sequence from the limit cycle) and eventually to regular spiking and to Shilnikov chaos, are to be expected.

In the case of Shilnikov chaos, the spread of the points in the return map defined near the homoclinic orbit (see Sections 3.3.3.b and 3.4.1) can be very small, and one has to consider another factor influencing the position of the points in the return map. This factor is the transient fluctuation enhancement, occurring when a macroscopic system decays from an unstable state (Arecchi, Gadomski et al., 1988; Arecchi, Lapucci et al., 1988), which was discovered many years ago by Arecchi et al. (1967 and 1980) (see also Haake, 1978). This enhancement occurs because of the fluctuation amplification properties of a system close to a stationary unstable point (this point has also been noted in Section 7.1). This amplification of relative fluctuations scales as the system size, $O(N)$ (i.e. it is proportional to the number of degrees of freedom N), so that it counterbalances the reduction in relative fluctuations brought about by the system size, which is $O(1/N)$ (i.e. the relative fluctuations decrease in proportion to $1/N$). Hence the amplified relative fluctuations will be $O(1)$ and the corresponding absolute fluctuations scale as the system size. This enhancement can be repeated at each cycle of the Poincaré return map. This effect is a specific indicator of intrinsic fluctuations and establishes a fundamental difference between chaotic experiments on large systems (i.e. real-life experiments) and simplified low-dimensional models (i.e. model simulation) (Arecchi, Gadomski et al., 1988; Arecchi, Lapucchi et al., 1988).

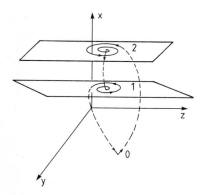

Fig. 7.2.3. Schematic view of a trajectory in the phase space when the dynamics is affected by all three unstable fixed points 0, 1 and 2. In our notation, $x = I$, $y = AD$ and $z = V$. The parameter values are as in Fig. 7.2.2: $A = 6.667$, $\alpha = 9.0$, $B = 0.838$ and $r = 0.16$. (From Arecchi, Gadomski et al., 1988.)

Swetits and Buoncristiani (1988) recently investigated the stability structure of eqns. (7.2.1) and, contrary to the interpretation above (Arecchi et al., 1986; Arecchi, Gadomski et al., 1988; Arecchi, Lapucci et al. 1988), no computational evidence of a Shilnikov-type instability for parameter values corresponding to the experimental conditions was found.

Chen et al. (1988) initiated a theoretical study of the stability and dynamic behavior of a CO_2 laser with feedback control of the cavity length. In this case the cavity detuning δ has to be taken into account, so that the starting equations are the set (7.1.14), to which one now has to add a third equation describing feedback of the laser intensity to control the cavity length L:

$$\dot{L}(t) = -\frac{1}{\tau}[L(t) - L_0 - \alpha I(t)] \qquad (7.2.3)$$

where τ is a Debye relaxation time, L_0 is the length of the laser without feedback and α is the strength of the feedback. In fact, it is simpler to write directly a differential equation for the cavity detuning instead of eq. (7.2.3). δ and L are related by

$$\delta \equiv (\omega_c - \omega_{ab})/\gamma_\perp = [2\pi m c/(2L) - \omega_{ab}]/\gamma_\perp \qquad (7.2.4)$$

where m is an integer and c is the spead of light, so that from eqns. (7.2.3) and (7.2.4) one obtains

$$\dot{\delta}(t) = -\frac{1}{\tau}[\delta(t) - \delta_0 + B I(t)] \qquad (7.2.5)$$

where $\delta_0 = [2\pi m c/(2L_0) - \omega_{ab}]/\gamma_\perp$ is the initial cavity detuning, $B = \omega_c \alpha/\gamma_\perp \cdot L_0$ and the approximation $L \approx L_0 \gg \lambda$ has been used.

Equations (7.1.14) and (7.2.5) form the final set required to describe the CO_2 laser with feedback control of the cavity length, and they are also approximately valid to describe other single-mode class B lasers such as the ruby, Nd:YAG, Nd:glass and semiconductor lasers.

The stationary solutions of eqns. (7.1.14) and (7.2.5) and their stability properties have been analysed theoretically by Chen and co-workers (1988). Bistability is found in the output intensity as the laser exitation is varied if the feedback is strong enough and if it has the correct sign to offset the initial cavity detuning. The stationary solutions can become unstable both in the presence or absence of bistable behavior. Unstable dynamical evolution is possible whenever the bandwidth of the feedback loop exceeds the relaxation rate of the population inversion, $\tau^{-1} > \gamma_\parallel$, because in this case the population is unable to follow the feedback fluctuations adiabatically. Under these conditions the instability threshold can be very low, and any small, yet non-zero, amount of feedback destabilizes the laser operation. Stability can be achieved only if $\tau^{-1} < \gamma_\parallel$ or if the bias is very large, $B \approx \kappa/\gamma_\parallel$ (in which case the intensity output is essentially zero). When operated under these conditions, the laser should show improved stability of both amplitude and frequency, i.e. the feedback loop can play its usual role of stabilizing the system.

In the middle of the unstable region the laser shows periodic or chaotic pulsations at a pulsation frequency $\omega_p \approx (\kappa \gamma_{\parallel})^{1/2}$, i.e. the relaxation frequency of the laser without feedback. The frequency varies as given by $(\gamma_{\parallel}/\tau)^{1/2} < \omega_p < \kappa$. Chen et al. (1988) have not yet determined if a Shilnikov instability associated with the point of the intermediate branch of stationary solutions (which has been described for the previous case) exists for the present kind of feedback. In any case, they never found conditions for the existence of three unstable stationary solutions.

7.2.2 Lasers with Optical Feedback

Optical feedback, i.e. a feedback loop consisting of an optical beam instead of an electronic current, can also be used with a laser. An example is the case of the re-injection of its spectrally filtered output field into a semiconductor laser. This is accomplished by passing the output field through a high-finesse resonator, as in Fig. 7.2.4. This kind of feedback has recently been used for frequency noise and laser line width reduction. Evidently it can also lead to dynamical instabilities.

A theoretical analysis of this kind of optical feedback in a single-mode semiconductor laser has been made by Li and Abraham (1988) using the differential-difference equations of Lang and Kobayashi (1980), including the addition of a feedback term for the external resonator, and Telle and Li (1989) using the model of Henry and Kazarinov (1986). Other related theoretical work has also been published (Henry, 1982 and 1983; Spano et al., 1984; Agrawal and Henry, 1988). In the first case (Li and Abraham, 1988; Lang and Kobayashi, 1980), the equations are

$$\frac{dE_0(t)}{dt} \frac{1}{E_0(t)} = \frac{1}{2} G_N \Delta N(t) - \frac{1}{2} G_I E_0^2(t) + \kappa \operatorname{Re}\left[\frac{E'(t)}{E(t)}\right] + F_\beta(t)$$

$$\frac{d\phi(t)}{dt} = \omega_0 - \omega + \frac{1}{2} G_N \Delta N(t)\alpha - \kappa \operatorname{Im}\left[\frac{E'(t)}{E(t)}\right] + F_\varphi(t) \qquad (7.2.6)$$

$$\frac{d\Delta N(t)}{dt} = \Delta J - \frac{\Delta N(t)}{\tau_s} - G E_0^2(t) + F_N(t)$$

where the slowly varying amplitude of the laser field is described by the amplitude E_0 and the phase ϕ, $G = G_0 + G_N \cdot \Delta N - G_I E_0^2$ is the net rate of stimulated emission $(G_N = \partial G/\partial N, G_I = \partial G/\partial I)$, ΔN represents the carrier number above the threshold value, ΔJ is the pump rate over the threshold, τ_s is the lifetime of the excited carriers, ω is the laser steady-state frequency, ω_0 is the laser frequency at the threshold, $\alpha = \Delta n'/\Delta n''$ is the phase-intensity coupling factor ($\Delta n'$ and $\Delta n''$ are the real and imaginary parts, respectively, of the deviations of the refractive index from the steady-state values),

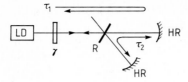

Fig. 7.2.4. Compound cavity configuration including a high-finesse resonator. γ is the field attenuation coefficient, τ_1 is the round-trip time between the LD and the external resonator, τ_2 is the round-trip time of the external resonator and R is the power reflectivity of the coupling mirror. (From Li and Abraham, 1988.)

$k = (1 - \gamma_{LD}^2)/(\gamma_{LD} \cdot \tau_{LD})$, γ_{LD} is the field reflectivity of the laser output mirror, τ_{LD} is the round-trip time of the laser diode cavity and $F_\beta(t)$, $F_\varphi(t)$ and $F_N(t)$ denote Langevin forces representing the spontaneous emission noise sources. The external resonator action is described by a function $E'(t)$ given by

$$E'(t) = \gamma(1 - R) \sum_{n=0}^{\infty} R^n E(t - \tau_1 - \tau_2 n) \tag{7.2.7}$$

where γ is the two-way field loss which accounts for the total field attenuation (feedback coefficient), τ_1 is the round-trip time between the laser diode and the external cavity, τ_2 is the round-trip time of the external resonator and R is the power reflectivity of the coupling mirror (which determines the finesse of the resonator). Hence the terms in eqns. (7.2.6) in which $E'(t)$ appears represent the re-injection of radiation from the external resonator into the laser cavity.

The external resonator can also be described by its response to different optical frequencies ω' through a complex transfer function (or frequency-dependent reflection coefficient), $T(\omega')$:

$$E'_{\omega'}(t) = E_{\omega'}(t) \cdot T(\omega') \tag{7.2.8}$$

with

$$T(\omega') = \gamma \frac{(1 - R)e^{i\omega'\tau_1}}{1 - Re^{i\omega'\tau_2}}. \tag{7.2.9}$$

The system (7.2.6) has been solved by Li and Abraham (1988) in the standard way for this kind of equation with stochastic noise sources. Figure 7.2.4 shows a typical experimental realization of the optical feedback scheme.

Figure 7.2.5 shows measured and calculated frequency noise spectra. The curves a_i are calculated for different feedback strengths and b_2 and b_3 represent measurements using two external resonators with different finesses.

The frequency noise is suppressed up to the Fourier frequency f_0 corresponding to the linewidth of the external resonator f_0. Above this level, the noise rises as is to be expected for a Lorentzian resonator linewidth, proportional to f^2. Curve b_1 for no

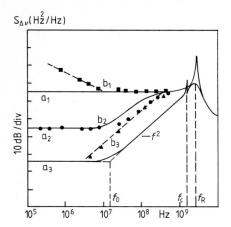

$S_{\Delta\nu}(\text{Hz}^2/\text{Hz})$

Fig. 7.2.5. Power spectral density of frequency noise. a_i (solid lines) are the calculated spectra, b_i (symbols) are the experimental results, a_1 is $\gamma = 0$ (free-running LD), a_2 is $\gamma = 0.01$. In the calculation $\tau_1 = \tau_2$. The external resonator has FSR $= 1.5$ GHz and the linewidth $2f_0 = 30$ MHz. The resonance frequency of external resonator ω_c is chosen to be equal to the free-running LD frequency. The feedback phase is chosen to keep the laser frequency with feedback equal to ω_c. b_1 is measured on a free-running LD, b_2 and b_3 are on laser diodes with different resonators. (From Li and Abraham, 1988.)

feedback shows the typically observed, but little understood, $1/f$ noise of the frequency fluctuations.

It must be noted that near the mode spacing frequency $f_c = c/2L$ of the external resonator, a dispersively shaped resonance occurs. Equally, the noise is enhanced at the relaxation oscillation frequency of the diode laser. Both of these resonances can, at higher feedback levels, result in pulsing through a supercritical Hopf bifurcation, as a stability analysis (Li and Abraham 1989) has shown.

7.2.3   Lasers with Electrical Feedback, Experiments

A particularly interesting feature of this kind of experimental study is that it permits the dynamics generated by several fixed points to be visualized, and that it is different from the Lorenz model. A schematic diagram of the experiment is shown in Fig. 7.2.1 (Arecchi et al., 1986 and 1987).

A linear stability analysis as mentioned above (Section 7.2.1) shows three fixed points: "0", the zero intensity solution, "1", which corresponds to the steady-state output in the absence of feedback, and a third point, "2".

"0" is a saddle point; it has three real eigenvalues, two negative and one positive, which means a two-dimensional stable and a one-dimensional unstable manifold (like the case of the Lorenz equations).

With increasing bias voltage B, "1" changes from a stable point to a limit cycle (Hopf bifurcation: after the bifurcation the eigenvalues are one real negative and two complex conjugate with a positive real part).

"2" has, for a wide range of B, a real positive eigenvalue and two complex conjugate eigenvalues with a negative real part (a sink focus saddle point: one-dimensional unstable and two-dimensional stable manifolds). With this information, the experimental observations can be interpreted.

Figure 7.2.6 shows the temporal behavior of the laser when B is increased. In case a) the point "1" is below the Hopf bifurcation: the laser emits in steady state. Case b) is after the Hopf bifurcation. Limit cycle behavior is seen, corresponding to small oscillations around the steady-state laser power. In case c) the limit cycle has widened so that it has come under the attracting influence of "0"; since the laser takes a long time to build up the field in the resonator, the laser remains a long time at "0"; it is then ejected along the unstable manifold direction of "0", loops around "2" owing to the attraction in the two-dimensional stable manifold of "2", is then weakly attracted by the steady state "1" and is finally attracted in the two-dimensional stable manifold of "0". In this case there is a clear influence of all three fixed points on the dynamics different from cases a) and b) where the "other" fixed points are too far away to influence the trajectories.

Figure 7.2.7 shows the temporal laser behavior at higher feedback gain when B is increased. In this case, the limit cycle formed after the Hopf bifurcation of "1" period doubles to reach chaotic motion locally around "1" which is shown in (a). Increasing B further then widens the chaotic attractor until it comes close to fixed point "2" (b), to which it is then attracted in the two-dimensional stable manifold in the spiral manner appropriate to the sink focus which "2" represents. The length and duration of the spiral obviously depend sensitively on the details of the approach towards "2", so that

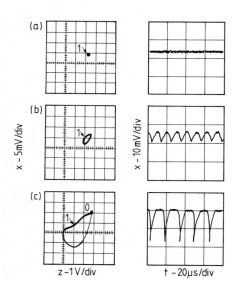

Fig. 7.2.6. Phase-space projections $z - x$ [(feeback voltage)-(laser intensity)] and time plots of the intensity $x(t)$ for low feedback gain. Intensity increases downward. (a) $B = 0.259$; (b) $B = 0.285$; (c) $B = 0.385$. Approximate locations in the phase plane of points 1 and 0 are indicated. (From Arecchi et al., 1987.)

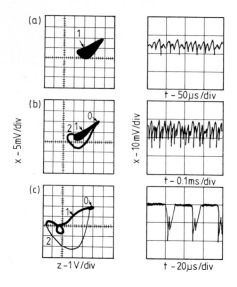

x – 5mV/div

x – 10mV/div

(a)

1

t – 50µs/div

(b)

0.
2 1

t – 0.1ms/div

(c)

0
1
2

z – 1V/div

t – 20µs/div

Fig. 7.2.7. Phase-space projections $z - x$ [(feedback voltage)-(laser intensity)] and time plots of the intensity $x(t)$ for high feedback gain. (a) $B = 0.296$; (b) $B = 0.311$; (c) $B = 0.411$. Also indicated are the approximate locations of points 0, 1 and 2 in the phase plane. (From Arecchi et al., 1987.)

chaotic dynamics result in a similar fashion to that of the focus points of the Lorenz model.

(c) As the trajectory expands further, it remains longer at the "0" point, which stabilizes the trajectory by damping out the fluctuations of the approach to "0", thus stabilizing the trajectory to a more periodic behavior.

It might be remarked that the dynamics of this laser with feedback resembles somewhat the laser with a saturable absorber (Chapter 6), where the steady state, the small limit cycle (period doubling) and the dynamics of the saddle focus in connection with the "0" saddle point are equally found, although with opposite signs of the eigenvalues and consequently time-reversed motion. The analogy can be understood from the interpretation of the saturable absorber as an intensity-dependent loss, like the intensity-driven electro-optic modular in the feedback laser.

7.3   Lasers with Injected Optical Fields

The effects of an injected signal on an oscillator were first studied and applied in electronic devices and masers. A first theoretical analysis of a laser with an injected optical field was performed by Spencer and Lamb (1972). Since then, progress has been made mainly in the cases of homogeneously broadened gas lasers (Bandy et al., 1985; Tredicce et al., 1985 and references cited therein), semiconductor lasers (Otsuka and Kawaguchi, 1984; Otsuka, 1985 and references cited therein) and the NMR laser have also been treated (Brun et al., 1985; Holzner et al., 1987).

The most important effect of an injected optical field on a laser is frequency locking, i.e. the laser amplification is driven by the injected field and emission occurs at the frequency of this field. Indeed, this constitutes a widely used technique for controlling the frequency of unstable powerful lasers from "master" lasers of high spectral purity or high-frequency stability (see, for instance, Weiss et al. (1980) for a class C laser). Frequency locking occurs for amplitudes of the injected field above a threshold value which increases with increasing detuning between the injected and laser fields. Below this threshold, the simultaneous presence of the injected and laser fields within the resonator leads to competition effects which result in dynamics.

7.3.1 Gas Lasers

If a monochromatic optical field is injected into the resonator of a homogeneously broadened single-mode uni-directional laser, an additional term has to be introduced into the first of eqns. (5.1.19), which then reads

$$\dot{E}(t) = -i\omega_c E(t) - \kappa [E(t) - \tilde{E}' e^{-i\omega' t}] + Ngp(t) \tag{7.3.1}$$

where $\tilde{E}' \exp(-i\omega' t)$ represents the injected field of amplitude $\tilde{E}'$ and frequency ω', and $E(t)$ now represents the total field inside the laser cavity. This total field couples through the second and third equations with the polarization and population inversion.

Lugiato and co-workers (Lugiato et al., 1983; Bandy et al., 1985) solved this set of five coupled real equations for several parameter settings for which no adiabatic elimination can be performed. They found regular and chaotic oscillations, a period-doubling cascade, envelope "breathing" and spiking. The basic features, however, can also be observed in the simpler case of a class B laser such as the CO_2 laser, where the polarization can be adiabatically eliminated and, in the general case of $\omega' \neq \omega_c, \omega_{ab}$ a system of only three real equations is obtained, i.e. sufficient degrees of freedom for chaotic behavior. Arecchi and co-workers (Arecchi et al., 1984a and b; Tredicce et al., 1985) found for this case the following set of equations which generalizes eqns. (7.1.3) when an injected signal is present and ω_c is different (in general) from ω_{ab}:

$$\dot{\tilde{E}}(t) = \tilde{E}(t)\left(-\kappa(1+i\theta) + \frac{Ng^2}{\gamma_\perp(1+i\delta)}d(t)\right) + \kappa\tilde{E}'$$

$$\dot{d}(t) = -\gamma_\parallel [d(t) - d_0] - \frac{4g^2}{\gamma_\perp(1+\delta^2)}|\tilde{E}(t)|^2 d(t) \tag{7.3.2}$$

where $\tilde{E}(t)$ in general is now complex and is defined by

$$E(t) = \tilde{E}(t)e^{-i\omega' t} \tag{7.3.3}$$

and θ and δ are given by

$$\theta = \frac{\omega_c - \omega'}{\kappa}, \quad \delta = \frac{\omega_{ab} - \omega'}{\gamma_\perp}. \tag{7.3.4}$$

Equations (7.3.2) can be put into real form by writing the complex field amplitude $\tilde{E}(t)$ as $\tilde{E}_R(t) + \tilde{E}_I(t)$. Furthermore, simplified expressions are obtained by substituting the dimensionless variables x, y and z, defined as:

$$x = \frac{2g}{(\gamma_\parallel \gamma_\perp)^{1/2}} \cdot \tilde{E}_R; \quad y = \frac{2g}{(\gamma_\parallel \gamma_\perp)^{1/2}} \tilde{E}_I; \quad z = \frac{Ng^2}{\kappa \gamma_\perp (1 + \delta^2)} d(t). \tag{7.3.5}$$

With this, eqns. (7.3.2) transform into

$$
\begin{aligned}
\dot{x} &= (z-1)x + (\theta - \delta z)y + A \\
\dot{y} &= -(\theta - \delta z)x + (z-1)y \\
\dot{z} &= \frac{\gamma_\parallel}{\kappa}\left(-\frac{x^2 + y^2}{1 + \delta^2} z - z + z_0 \right)
\end{aligned}
\tag{7.3.6}
$$

where $A = 2g\tilde{E}'/(\gamma_\parallel \cdot \gamma_\perp)^{1/2}$ ($\tilde{E}'$ is assumed to be real) and the dot now denotes the derivative with respect to the dimensionless time $\tau = \kappa t$.

The system (7.3.6) represents a set of three coupled real equations which yields stationary, periodic and chaotic solutions.

Inspection of eqns. (7.3.3) and (7.3.6) indicates that all steady-state solutions have frequency ω', i.e. they correspond to frequency locking of the laser to the injected signal. However, the steady-state solutions are stable only for amplitudes of the injected signal above a threshold value which can be zero for $\theta = \delta$ and increases with increasing $\theta - \delta$ (Tredicce et al., 1985). As $\theta - \delta$ is proportional to the difference between the injected and the "internal" laser frequency ω_L (i.e. the laser frequency corresponding to steady-state emission without injected field), $\theta - \delta = [(\kappa + \gamma_\perp)/\kappa\gamma_\perp](\omega' - \omega_L)$, this is in accordance with the predictions given above.

The general findings concerning stable and unstable behavior are summarized in the phase diagram in Fig. 7.3.1, which gives the areas of different dynamical properties. In the area marked "stable" the laser frequency is phase-locked to the externally

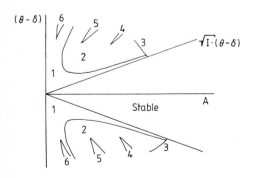

Fig. 7.3.1. State diagram for a CO_2 laser with an injected field in the detuning — external field amplitude — plane. For discussion, see text. (From Tredicce et al., 1985.)

injected field. The phase difference between laser and external signal is zero on the $\theta - \delta = 0$ line and approaches $\pm\pi$ at the stability boundaries. In the region marked "1" the system exhibits high-intensity pulses at low repetition frequency (Fig. 7.3.2a).

This is explained as follows. The external field locks the laser field for a relatively long period, during which the laser intensity is lower than for the free-running laser owing to the detuning by the externally induced frequency shift. The population inversion consequently rises until the unlocked laser emission exceeds the threshold. This results in a giant laser pulse which decreases the inversion abruptly (see Fig. 7.3.2c). During the giant pulse, the phase between the laser field and the external field "slips" by 2π. After this phase slip, the laser relocks to the external field (Fig. 7.3.2b). After the giant pulses, some ringing in the laser intensity occurs (see the inset in Fig. 7.3.2a). This is due to the relaxation oscillations following the large perturbation by the giant pulses. For large enough detuning of the external field from the laser field and sufficient external field strength (area 2), the frequency of the giant pulses can become comparable to the relaxation oscillation frequency. The relaxation oscillations then do not damp out completely between the giant pulses and become effectively undamped (Fig. 7.3.3a). If the time between the giant pulses is not approximately an integer number of the relaxation oscillation periods, "interference" between the giant pulses and the relaxation oscillation can lead to the appearance of subharmonics of the giant pulse frequency (area 4) (Fig. 7.3.3b).

When choosing a detuning and an injection field which makes the giant pulse frequency approximately equal to the relaxation oscillation frequency, one can expect a destabilization of the relaxation resonance similar to the case of modulated lasers.

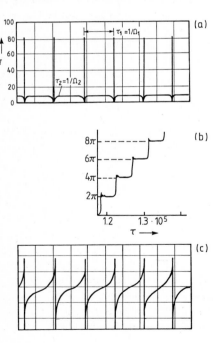

Fig. 7.3.2. Temporal evolution of the laser with injected field in region 1 of Fig. 7.3.1. (a) Laser output intensity showing giant (Q-switch-like) pulses; (b) phase difference between injected and laser field; each pulse corresponds to a 2π "phase-slip"; (c) medium inversion as a function of time. (From Tredicce et al., 1985.)

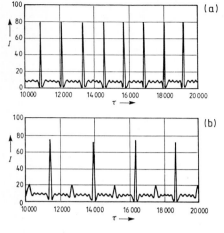

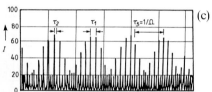

Fig. 7.3.3. Temporal evolution of the laser with injected field in region 2 of Fig. 7.3.1. There are four relaxation oscillations between large pulses in (a). $A = 0.0480$ the giant pulse frequency cannot lock with relaxation oscillation causing a period doubling (b). (From Tredicce et al., 1985.)

Fig. 7.3.4. When the pulsing frequency of the giant pulses becomes comparable to the relaxation oscillation frequency (region 5 in Fig. 7.3.1), the interaction of the two frequencies results in a low-frequency Ω_3. (From Tredicce et al., 1985.)

In area 5, as the relaxation oscillation frequency is approximately twice as high as the giant pulse frequency, a third low frequency appears (Fig. 7.3.4). When the field strength is then increased, the giant pulse frequency locks with the low frequency and the further pulsing shows an intermittent transition to chaos.

At the detuning and injection fields in area 6, a period-doubling transition to chaos (Fig. 7.3.5) is found. Generally, it is found that this system exhibits multiple coexisting solutions, so that the phase diagram (Fig. 7.3.1) is in effect superimposed with other phase diagrams and the various phases can be reached only by starting from different initial conditions ("hard-mode" excitation).

An experiment to test these predictions has been carried out by Boulnois et al. (1986). The laser under test is a ring laser emitting uni-directionally. The injection field is supplied by an additional CO_2 laser with active frequency stabilization. A chopper can interrupt the injection field so that the frequency difference between uninjected laser and the injection field can be measured. The output power of the laser under test is measured as a function of time. The frequency stability requirements on the two lasers are severe. The range in which the test laser is phase-locked to the injection signal is $1-3$ MHz at the available injection intensities. The dynamic phenomena to be measured are all predicted to occur within ca. 1 MHz away from the locking range.

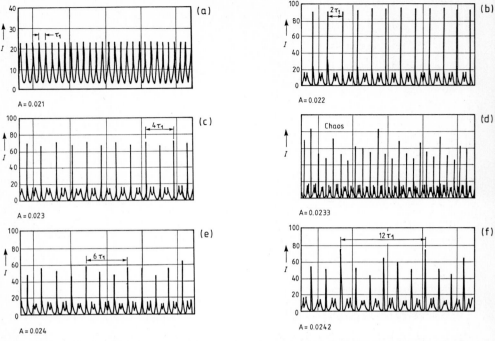

Fig. 7.3.5. Region 6 in Fig. 7.3.1. (a), (b), (c) and (d) show a period-doubling transition to chaos; (e) and (f) show a period-6 pulsing and its first subharmonic, period-12. (From Tredicce et al., 1985.)

Therefore, to conduct observations near the locking boundary, a frequency stability of a few times 10^{-10} is required.

Since the frequency of the injection field is stabilized for the measurements, the frequency of the injected laser was varied. Figure 7.3.6 shows the output power of the "injected" laser with a relatively high injection power (700 mW). At 1.5 MHz from the locking range boundary the sharp giant pulses are clearly observed. This corresponds to the situation in Fig. 7.3.6a. The pulsing frequency increases when the laser is tuned closer to the locking range boundary (Fig. 7.3.6b,c).

The relaxation oscillations caused by the large perturbation of the giant pulse are also observed in Fig. 7.3.7a, b, which corresponds to the situation in Fig. 7.3.6 with higher detuning. Clustering of pulses is finally observed (Fig. 7.3.8), possibly corresponding to the situation in Fig. 7.3.4.

Finally an irregular time dependence of the laser output is observed, possibly corresponding to the intermittent chaos of area 5. Therefore, qualitatively, the experiment shows the phenomena predicted.

Another possibility of achieving dynamics and chaotic behavior consists in injecting an optical field into a class A laser and simultaneously modulating one of the control

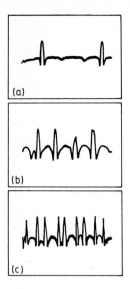

(a)

(b)

(c)

Fig. 7.3.6. Pulsing of injected CO_2 laser with three different detunings between injected and injection laser. Frequency difference: (a) 1500, (b) 600 and (c) 300 kHz. (From Boulnois et al., 1986.)

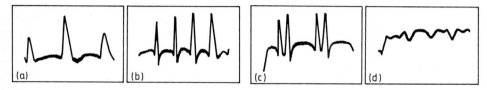

(a) (b) (c) (d)

Fig. 7.3.7. The giant pulses are accompanied by relaxation oscillations of small amplitude (a – d). (From Boulnois et al., 1986.)

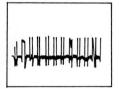

Fig. 7.3.8. Giant pulses of the injected laser appear in pairs. (From Boulnois et al., 1986.)

parameters, e.g. the injected field, the pump rate or the cavity losses, as has been theoretically analyzed by Yamada and co-workers (Yamada and Graham, 1980; Scholz et al., 1981).

7.3.2 Semiconductor Lasers

Since injection locking is of substantial practical importance for optical communication systems, it has been widely investigated and its dynamical properties have been studied.

A distinctive feature of semiconductor lasers compared with other lasers is the strong dependence of the active layer refractive index on carrier or population-inversion density. This results in a dependence of the gain and laser frequency on carrier density which strongly influences the amplification dynamics because the carrier density is depleted by the injection of the external field and by the generation of the laser field. Therefore, this system can be considered as a non-linear Fabry-Pérot interferometer with true optical gain and non-linear refraction (gain-linked dispersive optical system).

Otsuka and co-workers (Otsuka and Iwamura, 1983; Otsuka and Kawaguchi, 1984a and b; Otsuka 1985) studied theoretically semiconductor lasers with an injected field in several configurations. In some studies (Otsuka and Iwamura, 1983; Otsuka and Kawaguchi, 1984a; Otsuka, 1985) a laser amplifier with an external mirror working under resonant conditions was considered, whereas in (Otsuka and Kawaguchi 1984b) a simpler laser resonator consisting of the two parallel facets of the semiconductor diode material and working under detuned conditions was dealt with. Let us briefly describe this last case, where the dynamic behavior is related to the anomalous dispersion effect for frequencies close to the maximum-gain frequency.

Assuming negligible standing-wave spatial modulation of the population inversion (spatial hole burning) and propagation effects, and employing the uniform-field approximation, the equations describing this kind of laser have been written as (Otsuka and Kawaguchi, 1984b):

$$\frac{\mathrm{d}\tilde{E}}{\mathrm{d}t} = \frac{\tilde{E}}{2}\left[-\frac{1}{\tau_\mathrm{p}} + g(n_\mathrm{e})\right] + \frac{\omega_0}{2Q}\tilde{E}'\cos\Psi$$

$$\frac{\mathrm{d}\Psi}{\mathrm{d}t} = [\omega_0(n_\mathrm{e}) - \omega'] - \frac{\omega_0}{2Q}\frac{\tilde{E}'}{\tilde{E}}\sin\Psi \qquad (7.3.7)$$

$$\frac{\mathrm{d}n_\mathrm{e}}{\mathrm{d}t} = P - \frac{n_\mathrm{e}}{\tau_\mathrm{e}} - g(n_\mathrm{e})\tilde{E}^2$$

where E represents the electric field amplitude, $\tilde{E}^2 \equiv n_\mathrm{p}$ is the photon density, τ_p is the photon lifetime ($\tau_p^{-1}/2 \equiv \kappa$), n_e is the population-inversion density, $g(n_\mathrm{e})$ is the gain function, $\omega_0(n_\mathrm{e})$ is the laser frequency, $\tilde{E}'$ and ω' are the injected-field amplitude and frequency respectively, Ψ is the (external) phase angle, Q is the (external) quality factor of the resonator, P is the pump rate and τ_e is the upper-state lifetime. Note that in contrast to previous cases, such as eqns. (6.1.5) and (7.1.16), in eqns. (7.3.7) one has to deal with the field amplitude $\tilde{E}$ and phase Ψ (instead of only the photon density) in order to take into account detuning and dispersion effects.

As we have indicated, the dependences of g and ω_0 on n_e arise because the active-region refractive index varies with the population-inversion density as a result of the anomalous dispersion in the detuned laser medium. For brevity, $g(n_\mathrm{e})$ and $\omega_0(n_\mathrm{e})$ were approximated using Taylor's series up to the first order (Otsuka and Kawaguchi, 1984b):

$$g(n_e) = g(n_e^{th}) + \frac{\partial g}{\partial n_e}(n_e - n_e^{th}) = \frac{1}{n_e^{th}\tau_p} + \frac{\partial g}{\partial n_e}(n_e - n_e^{th})$$

$$\omega_0(n_e, \Omega) = \omega_0(n_e^{th}) + \frac{\partial \omega_0}{\partial n_e}(n_e - n_e^{th}) - \left(\frac{N_e}{N} - 1\right)[\Omega - \omega_0(n_e^{th})] \qquad (7.3.8)$$

where n_e^{th} is the threshold population density and the dependence on the actual oscillation frequency Ω brought about by the anomalous dispersion has been included in the last equation (N_e and N denote refractive indices: $N_e = N + \Omega(\partial N/\partial \Omega)$ is the effective refractive index). A parameter $R = -2(\partial\omega_0/\partial n_e)/(\partial g/\partial n_e)$ can be defined, which in general takes on values between -0.5 and -6.2. We also define further dimensionless parameters: $w = P/P_{th}$ (P_{th} = threshold pump rate), the normalized pump rate; $S_0 = \tilde{E}^2/\tau_P P_{th}$, the normalized photon density; $S_i = (\omega_0 \tilde{E}'/2Q)^2 \tau_p/P_{th}$, the normalized injection photon density; and $\Delta\omega = (N_e/N)[\omega' - \omega_0(n_e^{th})]\tau_p$, the normalized frequency detuning.

The findings in the numerical investigations are summarized as for the CO_2 laser in a phase diagram (Fig. 7.3.9) in the detuning, pump strength plane. In the bistable region I, the intensity depends on the direction of tuning of the injection field frequency. In region II the laser is phase-locked to the injection field with the phase difference approaching $\pm\pi$ at the boundaries of the (stable) locking region.

At the pump rate 0.5 (right-hand side of Fig. 7.3.9) the laser dynamics is governed by period doubling: Fig. 7.3.10 shows different intensity pulsing patterns as the detuning is increased. Figure 7.3.10a falls in the stable region II. A steady state is reached after damped relaxation oscillations. Figure 7.3.10b shows undamped relaxation oscillations; c, d, e (region III) show the period-doubling transition to chaos; in f (region IV) the

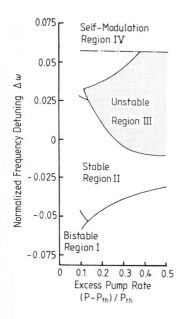

Fig. 7.3.9. State diagram for a semiconductor laser with injected field. P = pump strength; P_{th} = pump strength at first laser threshold; $\Delta\omega$ = detuning between laser and injection field frequency normalized to the laser resonator linewidth. (From Otsuka and Kawaguchi, 1984a.)

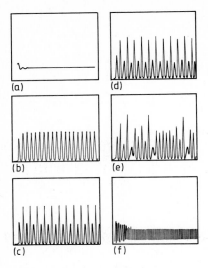

Fig. 7.3.10. Intensity pulsing for a semiconductor laser with injected field as calculated for $P = 1.5 P_{th}$ and varying detuning. (a) (locked) $\Delta\omega = -0.025$; (b) (period 1) $\Delta\omega = 0$; (c) (period 2) $\Delta\omega = 0.0125$; (d) (period 4) $\Delta\omega = 0.0156$; (e) (chaotic) $\Delta\omega = 0.0188$; (f) (non-interacting) $\Delta\omega = 0.0625$. (From Otsuka and Kawaguchi, 1984a.)

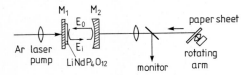

Fig. 7.3.11. Set-up to observe laser instabilities with external field injection. M = mirrors; $LiNdP_4O_{12}$ solid-state laser material pumped by Ar^+ laser (E, electric field amplitude). (From Otsuka and Kawaguchi, 1984a.)

two fields no longer interact substantially so that essentially the beating between the two fields is seen.

It might be expected that at the boundary between regions III and IV intermittent onset of chaos would occur, in analogy with the CO_2 laser.

An experimental indication that period doubling to chaos occurs under these conditions was obtained by using an optically pumped $LiNdP_4O_{12}$ laser instead of a diode laser. The system used is shown in Fig. 7.3.11. The pump is a 514 nm Ar^+ laser. The laser crystal is mounted directly on a flat mirror. The other mirror is concave to form an almost folded-confocal resonator. This arrangement ensures single-mode emission by avoiding "spatial hole burning" in the crystal. At the same time, the crystal acts as a Fabry-Pérot interferometer to select particular frequencies, detuned from the gain line center to obtain the desired dn/dI. As the detuning of the injected light from the laser frequency had to be smaller than the laser frequency fluctuations, frequency-shifted light from the laser itself was fed back into the laser. This was accomplished by scattering the laser light from a rotating inclined paper surface (see Fig. 7.3.11). The frequency shift of the scattered light is brought about by the Doppler effect. The output power spectrum of the laser then showed in fact a period doubling (Fig. 7.3.12a) and

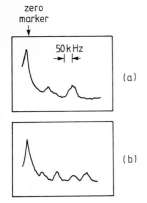

zero
marker

50 kHz

(a)

(b)

Fig. 7.3.12. Intensity spectrum of a $LiNdP_4O_{12}$ laser with injected signal as shown in Fig. 7.3.11 showing period doubling: (a) fundamental pulsing frequency and subharmonic; (b) subharmonic and second subharmonic. (From Otsuka and Kawaguchi, 1984a.)

period quadrupling (Fig. 7.3.12b) with increasing detuning, qualitatively in accordance with the predictions.

In interpreting the experimental observations with the laser-with-injected-signal equations, some caution is advisable. The scattered light (i) is incoherent in some sense and (ii) represents a delayed (although frequency-shifted) feedback. Whether the observed instabilities do not correspond to an Ikeda-like instability is not clear.

The asymmetry of the phase diagram with respect to detuning between the laser and injection field is due to the detuning of the laser with respect to the gain line center. This, of course, changes the sign (+ or −) of dn/dI, which is the cause of the asymmetry.

Morozov (1988) summarized the theoretical results on semiconductor injection lasers reported by Soviet researchers.

7.3.3   NMR Laser

As mentioned in Section 7.1, a nuclear magnetic resonance (NMR) laser (Brun et al., 1985) can be described by a set of equations equivalent to eqns. (5.1.19), so that when a driving oscillating magnetic field is injected the first of these equations transforms, as in the case of gas lasers considered in Section 7.3.1, into eq. (7.3.1).

Hence the theoretical treatment of Lugiato and co-workers (Lugiato et al., 1983; Brandy et al., 1985) could in principle be applied to NMR lasers. The only difference is that, as mentioned in Section 7.1, for an NMR laser one has $\kappa \approx 10\gamma_\perp \approx 4 \cdot 10^4 \gamma_\parallel$, so that it operates under conditions close to adiabatic elimination of the field. If the NMR laser is tuned to resonance and the injected field is detuned, a system of three real equations (two equations for the polarization and one for the population inversion) is obtained when the field is adiabatically eliminated.

The theoretical results (Brun et al., 1985) obtained with this model display qualitatively the same behavior as those for the CO_2 laser described above. Figure 7.3.13 shows the pulsing as calculated for the conditions of the NMR laser. The giant pulses

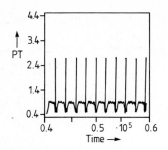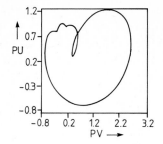

Fig. 7.3.13. Intensity pulsing of an NMR laser with injected field and corresponding phase portrait. (From Brun et al., 1985; Brun et al., 1986.)

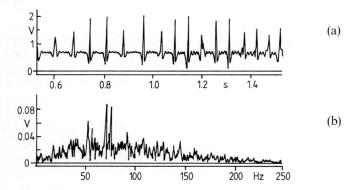

Fig. 7.3.14. Chaotic pulsing of the NMR laser near the injection locking point. (a) Time picture; (b) intensity spectrum. (From Brun et al., 1986.)

and the relaxation oscillations are recognizable. Figure 7.3.14 shows irregular pulsing measured near the injection-locking point. The spectrum is noise-like and suggests chaotic pulsing.

The NMR laser with an injected field has also allowed the unambiguous observation for the first time in physical reality of a subcritical Hopf bifurcation (Holzner et al., 1987). In contrast to a supercritical Hopf bifurcation for which a stable fixed point bifurcates to a stable limit cycle (periodic solution) lying around an unstable fixed point, for the subcritical Hopf bifurcation an unstable fixed point bifurcates to an unstable limit cycle lying around a stable fixed point (Fig. 7.3.15).

This makes the observation of a subcritical Hopf bifurcation more complicated than that of a supercritical one, since an unstable limit cycle has to be located. Theoretically this can easily be reached by calculating time-reversed solutions of the differential equations, but experimentally this trick does not work and the only way is through backward extrapolation of transient behaviors near the presumed unstable cycle. In the NMR laser experiment the pump is first set to the range of the stable fixed point above the bifurcation point. Setting the initial conditions so that the system is within

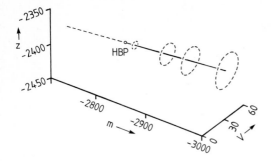

Fig. 7.3.15. Stable and unstable fixed points (solid and dashed straight lines) near the subcritical Hopf bifurcation point. Unstable limit cycles are shown (m: modulation amplitude, z, v: components of the laser field). (From Holzner et al., 1987.)

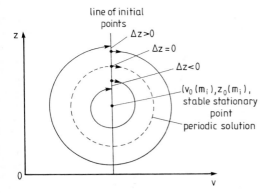

Fig. 7.3.16. Transient behavior of the system in the presence of an unstable limit cycle. Trajectories initiating outside the limit cycle are repelled towards infinity, inside towards the fixed point (z, v: components of the laser field, m: modulation amplitude). (From Holzner et al., 1987.)

the unstable limit cycle then results in a damped oscillation as shown in Fig. 7.3.16. Setting the initial conditions so that the system is outside the limit cycle results in a growing oscillation (Fig. 7.3.16). The corresponding measured oscillations are shown in Fig. 7.3.17a, b. These two different behaviors for the same parameters but different initial conditions show the existence of an unstable limit cycle. The bifurcation was then explored in detail by changing the distance to the bifurcation point, i.e. the pump strength.

We mention here that another possible method for detecting a subcritical Hopf bifurcation would be given by the Lorenz model. At a pump strength close to the onset of chaos, the stable chaotic (Lorenz) attractor coexists with two stable fixed points corresponding to the steady-state laser emission (see Section 2.1; for instance, Figs. 2.1.3a and 2.1.4a). The basins of attraction of the fixed points in the centers of the Lorenz attractor are separated from the basin of attraction of the chaotic Lorenz attractor by an unstable limit cycle. If the initial conditions, e.g. the field of the laser,

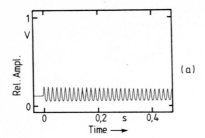 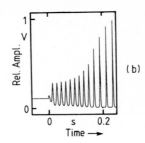

Fig. 7.3.17. Observed transient behavior of an NMR laser corresponding to the two cases in Fig. 7.3.16. (a) Transient originating inside the limit cycle (damped); (b) transient originating outside the limit cycle (undamped). (From Holzner et al., 1987.)

can be set outside the attractor (this is technically possible with far-infrared lasers) the motion will be chaotic. Inside the unstable limit cycle, the motion will be a damped oscillation.

The particular structure of the Lorenz attractor with its jumps from one leaf to another offers an easy way to jump into the interior of the unstable limit cycle. For a pump strength $r < r_A$ (this occurs, for instance, in the case of Fig. 2.1.3), where the chaotic attractor has not yet settled down, and since the point to which a trajectory jumps on one leaf is random, the system, after a period of erratic motion, will jump to a point inside the unstable limit cycle and will then be attracted to zero ("transient chaos"). Unlike the case of the subcritical Hopf bifurcation in the NMR laser with an injected signal, in the Lorenz model for $r < r_A$ it is therefore unnecessary to set artificially the initial condition inside the unstable limit cycle; the system will be trapped there anyway after some time. However, for $r_A < r < r_H$ (see Section 2.1), where the chaotic attractor has settled down, it is also necessary to set the initial condition inside the unstable limit cycle (or, more precisely, within the basin of attraction of the stable fixed points).

In contrast to the Lorenz model where the attractor outside the unstable limit cycle is a chaotic one, in the case of the injected NMR laser it is a periodic one.

8 Three-Level Optically Pumped Single-Mode Gas Lasers

8.1 General Considerations

8.1.1 Introduction

Of the lasers discussed in the preceding chapters, the amplifying medium can be described by a simple two-level model in which pumping is incoherent, such as electrical discharge excitation. A different kind of laser uses coherent light for pumping, which acts on a coupled transition sharing the upper level with the laser transition (Fig. 8.1.1).

The difference from the previous cases is that the coherence of the pumping field creates coherences in the medium and new phenomena, influencing the laser amplification process, may occur. For instance, the coherent dipole oscillation induced by the pump field on the $2-0$ transition can couple non-linearly with the dipole oscillation existing on the $0-1$ transition. Raman pumping may occur, with which amplification can occur without population inversion between levels 0 and 1. Similarly, the AC Stark splitting induced by the pump field on the common level "0" may split the gain profile for the laser field.

The theoretical description of the laser dynamics allowing for these additional phenomena requires a three-level model in which the interaction with both pump and generated field has to be included (Fig. 8.1.1). This leads, in general, to a larger set of coupled equations than in the previous cases of two-level systems, which is in principle more difficult to handle. On the other hand, the presence of the above-mentioned coherence phenomena can lead to a richer dynamic behavior and low instability thresholds.

As with two-level lasers, different types of broadening mechanisms (homogeneous and inhomogeneous) occur and different resonators (ring or Fabry-Pérot) are used in optically pumped lasers. Conditions allowing for adiabatic elimination of some of the dynamical variables are only exceptionally met in the optically pumped lasers investigated so far. Interestingly, conditions have been found for which optically pumped far-infrared gas lasers show experimentally a behavior of the type predicted by the Lorenz model (see Chapter 5). In fact, up to now these lasers constitute the only physical systems in which such dynamics has been observed.

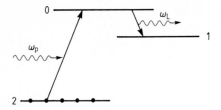

Fig. 8.1.1. Three-level system showing the coupling between the optical fields and the molecular medium in an optically pumped gas laser. A pump beam of frequency ω_P acts on the $2-0$ molecular transition and a laser beam of frequency ω_L is generated at the adjacent transition $0-1$.

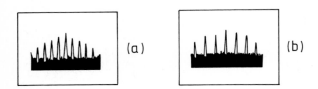

Fig. 8.1.2. (a) Frequency spectrum of the $^{13}CH_3F$ harmonic mixing experiment. The oscillation frequency at 0.007 Torr is 1.0 MHz. (b) Oscillation frequency at 0.015 Torr = 1.5 MHz. (From Lawandy et al., 1980.)

8.1.2 General Aspects

In the past, using the technique of optically pumping gases, a large number of laser emission lines (many thousands) have been excited (for a review, see Wellegehausen, 1979). Of most practical importance are laser-pumped gas lasers emitting in the far-infrared region. These emit on rotational molecular transitions. Similarly, vibrational transitions can be made to lase in the mid-infrared, and electronic molecular transitions in the near-infrared and visible spectral ranges. These lasers typically operate at low pressure, which implies small pressure broadening rates and consequently in the absence of spontaneous emission narrow homogeneous linewidths. This offers the possibility of fulfilling the bad-cavity condition (2.1.7) or (5.2.8): $\kappa > \gamma_\| + \gamma_\perp$, at least for some of those lasers without excessive laser resonator loss. For this reason, optically pumped gas lasers have been suggested as systems for studying single-mode laser instabilities (Weiss and Klische, 1984).

In fact, undamped pulsing had already been observed in harmonic mixing experiments in which far-infrared optically pumped lasers were used (Lawandy and Koepf, 1980). In these experiments, the optical spectrum of the laser output — observed as the radio-frequency spectrum of the mixing signal between the laser and a millimeter-wave harmonic — showed a series of equally spaced frequencies not compatible with the laser resonator mode frequencies (Fig. 8.1.2). Therefore, they had to be interpreted as a laser pulsing frequency and its harmonics. These are expected if the laser pulses coherently, i.e. the optical phase in each pulse is related to the optical phase of the preceding pulse.

The onset of undamped pulsing was properly ascribed to an instability of the steady-state emission, but erroneously as an instability of the usual class B laser rate equations, which do not have undamped pulsing solutions (Abraham et al., 1983).

As we have indicated above, one direct consequence of the coherent pumping of the pump transition is a splitting of the laser gain line (Fig. 8.1.3). In the presence of a strong pump field, the energy levels 2 and 0 (Fig. 8.1.1) split. Correspondingly, there are two possible laser transitions from level 0 to 1. They are separated in frequency by the Rabi frequency of the pump transition, 2β, where β is given by

$$\beta = \frac{\mu_{02} E_{\mathrm{p}}}{2\hbar} \tag{8.1.1}$$

(μ_{02} denotes the pump transition dipole moment and E_{p} the pump field strength). Physically, the two laser transitions become separated in frequency when the pump Rabi frequency exceeds the homogeneous linewidth, which means roughly when the pump transition becomes saturated. The gain line then has two maxima. Understandably, the laser will then behave dynamically differently from a simple inverted two-level system having a single (Lorentzian) gain line.

Figure 8.1.4 shows gain line shapes as calculated for a popular far-infrared gas laser: at high pump power levels the gain line has two maxima. Also shown is the gain saturation: the splitting disappears for higher strength of the generated laser field (Heppner and Hübner, 1980). This saturation behavior, however, is not to be confused with the gain that a weak probe field would see in the presence of the pump field and the strong laser field. This probe gain determines the amplification of fields of frequencies other than that of the generated laser field, and is thus relevant to the possibility of oscillatiory instabilities (i.e. spontaneous pulsing of the laser). Like the two-fold splitting of the levels in the presence of one field (i.e. the pump field as mentioned above), the presence of two strong fields (i.e. the pump field and the generated laser field) will split all three levels of the system three-fold. The result is that the probe gain

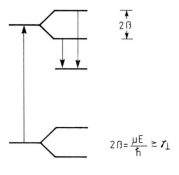

Optical (AC) Stark-Effekt

Fig. 8.1.3. Optical (AC) Stark effect; the levels connected by a strong field are split by an amount equal to the Rabi frequency.

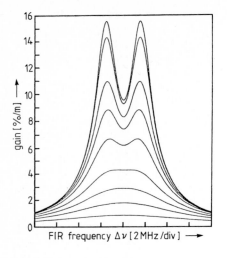

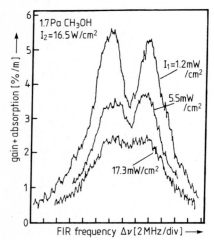

Fig. 8.1.4. AC Stark splitting of the gain line calculated for a typical laser-pumped gas laser (CH_3OH). Gas pressure, 5 Pa; pump field intensity, 15 W/cm^2. Laser (FIR) intensities: 0.1, 1, 5, 20, 50, 100, 200 and 500 mW/cm^2 (lowest curve: highest intensity). (From Heppner and Hübner, 1980.)

Fig. 8.1.5. Measured AC Stark splitting of the laser in Fig. 8.1.4. (From Heppner and Hübner, 1980.)

line is split five-fold and thus has a considerably more complicated structure than the saturated gain line (Fig. 8.1.4).

Measurements on optically pumped lasers have shown that these calculations correspond to reality. Compare Fig. 8.1.5 with Fig. 8.1.4, which shows a gain measurement corresponding to the conditions of Fig. 8.1.4. Together with the correct gain line shapes, measurements of this kind have also shown that calculations of the coherent interaction of two fields with a three-level system yield the proper gain values. Figures 8.1.6 and 8.1.7 show a comparison of measured and calculated gain for an ammonia laser far-infrared transition. The small signal gain and the saturation case show good agreement between the measured and calculated gain as a function of working pressure.

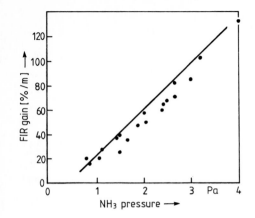

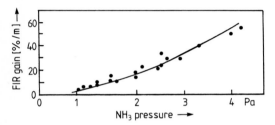

Fig. 8.1.6. Comparison of unsaturated gain as function of laser working pressure for the 81 μm ammonia laser, measurement vs. calculation. (From Marx et al., 1981.)

Fig. 8.1.7. Comparison of saturated gain as function of laser working pressure for the 81 μm ammonia laser, measurement vs. calculation. (From Marx et al., 1981).

It appears, therefore, that for these optically pumped lasers realistic theoretical models can be constructed. According to eqn. (5.1.21), the optically pumped three-level gas laser is described by the following set of equations if traveling waves, spatially uniform materials and fields and homogeneous broadening are assumed (Dupertuis et al., 1986; Pujol et al., 1988):

$$\dot{\varrho}_{00} = \gamma_0(\varrho_{00}^0 - \varrho_{00}) + 2\alpha\,\text{Im}\,(\varrho_{01}) + 2\beta\,\text{Im}\,(\varrho_{02})$$

$$\dot{\varrho}_{11} = \gamma_1(\varrho_{11}^0 - \varrho_{11}) - 2\alpha\,\text{Im}\,(\varrho_{01})$$

$$\dot{\varrho}_{22} = \gamma_2(\varrho_{22}^0 - \varrho_{22}) - 2\beta\,\text{Im}\,(\varrho_{02})$$

$$\dot{\varrho}_{01} = -\gamma_{01}\varrho_{01} + i\Delta_1\varrho_{01} + i\beta\varrho_{12}^* - i\alpha(\varrho_{00} - \varrho_{11})$$

$$\dot{\varrho}_{02} = -\gamma_{02}\varrho_{02} + i\Delta_2\varrho_{02} + i\alpha\varrho_{12} - i\beta(\varrho_{00} - \varrho_{22}) \qquad (8.1.2)$$

$$\dot{\varrho}_{12} = -\gamma_{12}\varrho_{12} - i(\Delta_1 - \Delta_2)\varrho_{12} + i\alpha\varrho_{02} - i\beta\varrho_{01}^*$$

$$\dot{\alpha} = -\kappa\alpha + G\,\text{Im}\,(\varrho_{01}^*)/2$$

where ϱ_{ij} $(i,j = 0,1,2)$ represent the elements of the 3×3 density matrix normalized to the density N^0 of molecules in the three-level system (ϱ_{ii} describes the population of level i, ϱ_{ii}^0 is the corresponding zero-field population, ϱ_{02} is the pump-induced coherence, ϱ_{01} is the coherence related to the laser field and ϱ_{12} is the Raman coherence), γ_i, γ_{ij} and κ are the population, coherence and field decay rates, respectively, $2\alpha = \mu_{01} E_L/\hbar$ is the laser-field Rabi frequency (μ_{01} is the laser-transition dipole moment and E_L is the laser field strength), β is defined by eq. (8.1.1) and is assumed to be constant, $\Delta_2 = \omega_P - \omega_{02}$ is the pump detuning with respect to the $2-0$ transition, $\Delta_1 = \omega_L - \omega_{01}$ is the laser field detuning with respect to the $0-1$ transition and $G = \omega_{01} \mu_{01}^2 N^0/\varepsilon_0 \hbar$ is the gain parameter. The fast oscillating factors $\exp(-i\omega_P t)$ and $\exp(-i\omega_L t)$, which ϱ_{02} and ϱ_{01}, respectively, contain, have been omitted, so that all variables of eqns. (8.1.2) represent slowly varying quantities. Note that the gain is determined by two parameters: G, which is mainly related to the molecular density and the laser transition's characteristics, and β, which describes the pump field amplitude and its coupling with the pump transition.

The actual laser frequency ω_L or, equivalently, the laser detuning Δ_1, is in general influenced by pulling and pushing effects and, therefore, is time dependent if the laser exhibits dynamics. The instantaneous value of Δ_1 is given by

$$\Delta_1 = \Delta_1^c - G\,\mathrm{Re}(\varrho_{01})/2\alpha \tag{8.1.3}$$

where $\Delta_1^c = \omega_c - \omega_{01}$ is the empty-cavity detuning. The time integration of Δ_1 provides the time dependence of the phase of the laser field.

Since ϱ_{01}, ϱ_{02} and ϱ_{12} are, in general, complex quantities, eqns. (8.1.2) represent a set of ten coupled real equations. Under resonant conditions ($\Delta_2 = \Delta_1^c = 0$), one has $\mathrm{Re}(\varrho_{02}) = \mathrm{Im}(\varrho_{21}) = 0$, so that the system reduces to seven coupled equations. Thus, the number of degrees of freedom is larger than in the case of a two-level laser. Dupertuis et al. (1986) showed that eqns. (8.1.2) transform to the Lorenz equations (5.2.6) or (2.1.1) describing an incoherently pumped homogeneously broadened resonantly tuned two-level laser in the following limit:

(a) the fields are resonant: $\Delta_1^c = \Delta_1 = \Delta_2 = 0$;
(b) $\gamma_0 = \gamma_1$;
(c) $\varrho_0^0 = \varrho_1^0$;
(d) the pump is weak (below saturation);
(e) γ_{02}, $\gamma_{12} \gg \gamma_0$, γ_{01};
(f) $\gamma_{02} \cdot \gamma_{12} > \alpha^2$;

where conditions (a) and (b) are already required in the two-level case. Condition (e), which allows adiabatic elimination of ϱ_{02} and ϱ_{12}, is required to neglect pump-induced coherent effects. Condition (a) can, of course, always be fulfilled. Condition (d) was estimated to yield sufficient gain to reach the second laser threshold (Weiss and Klische, 1984). No estimate is possible to show if (e) and (f) can be fulfilled in a real laser.

If the homogeneous linewidth is smaller than the Doppler width in an optically pumped gas laser, the different velocity classes of active molecules contribute to the laser gain in different ways through their resonant and Raman emission. Every velocity class yields a gain line with two maxima: one at the frequency corresponding to its Doppler shift on the laser transition and the other at a frequency offset from the first given by the generalized Rabi frequency associated with the pump transition $\sqrt{(2\beta)^2 + (k_P v)^2}$, where k_P is the pump field wavenumber, v is the molecular velocity and $k_P \cdot v$ the Doppler shift. To obtain the gain line profile formed by all molecules, the gain has to be integrated over all velocities (or Doppler shifts). The result now depends on whether the pump field and generated field are co- or counter-propagating, since the two cases differ in the sign of the Doppler shifts. The integration therefore yields different line profiles for the two cases (Fig. 8.1.8).

For forward emission (co-propagating fields), the gain line is split; Fig. 8.1.8 actually corresponds to the case where all molecules are taken into account. For backward emission (counter-propagating fields), the gain line has only one maximum. The shape of this line is neither Gaussian nor Lorentzian. If one wants to avoid the AC Stark splittings in these three-level gas lasers, it appears intuitively favorable to use the backward emission since it would seem to be more difficult to broaden away the split structure of the forward gain line than to make the backward gain line similar to a Lorentzian, which it resembles even in the presence of 3-level coherence effects.

Even though it appears possible to operate optically pumped gas lasers in the regime where they correspond to the Lorenz model, in general, however, one would expect laser dynamics strongly influenced by the coherence effects — different from the predictions of the Lorenz model.

Theoretical models of these lasers have so far not taken into account magnetic sublevel degeneracy. A molecule with total angular momentum J has $2J + 1$ energetically degenerate magnetic substates. If the pump and generated field polarizations are the same, then the optical transitions do not change the magnetic quantum number. In this case the different magnetic sublevels, which have different dipole moments with respect to the polarized fields, can be treated independently. Their transition rates can be added. If the polarizations of the pump field and generated laser field are mutually

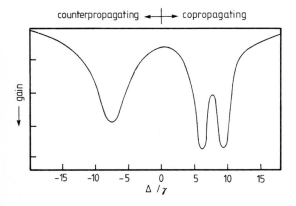

Fig. 8.1.8. Gain of an optically pumped laser for pump co-propagating with laser field (positive Δ) and counter-propagating to laser field (negative Δ).

perpendicular, the one field can be decomposed with respect to the other one into two circularly counter-rotating polarized waves. These circular components exert a torque on the molecular dipoles and the transitions then change the magnetic quantum numbers by ± 1.

In a strongly saturating laser field, the different states will all be connected by optical transitions. In this case the independent treatment is obviously not possible; rather, instead of a 3-level system a $(2J + 1) \cdot$ 3-level system has to be treated. Owing to the immense computational work necessary, this has not yet been attempted. It is therefore unknown what the influence of crossed polarizations is on the laser dynamics. In the following sections we describe the specific cases that have been investigated so far.

8.2   Cw Far-Infrared Lasers

8.2.1   Uni-Directional Ring Lasers

Most of the investigations on optically pumped lasers have been performed for the case of cw uni-directional emission in the far-infrared region. Let us first describe the experimental results and then the theoretical analyses.

a. Experimental Results

Most investigations have been performed on the aR(7,7) 81 µm emission line of the $^{14}NH_3$ laser (pumped by a ca. 10 µm N_2O laser line), and on the aR(4,4) 153 µm and aR(2,0) 374 µm emission lines of the $^{15}NH_3$ laser (pumped by a CO_2 laser). The first two cases have in common that the rotational quantum number K is different from zero, so the rotational levels involved are inversion doublets. In the third case, $K = 0$, so that only one of the levels of the doublet exists. As a result, the values of the molecular relaxation parameters can be different in both cases, which may be the reason for the different dynamic behavior, described below.

As described in Chapter 5 for the 81 µm ammonia laser, the dynamical properties were found to agree with the predictions of the Lorenz model (homogeneously broadened two-level laser) for parameter ranges where coherence effects of the pump are least likely. However, largely for low pressures where pump coherence effects are likely to be important, differences from the Lorenz model are found experimentally: (a) the pulse shapes are no longer spirals (Fig. 8.2.1); (b) the onset of chaotic emission proceeds via period doubling when the pump strength is increased, instead of the abrupt onset characteristic for the Lorenz model; (c) the threshold for chaos is substantially lower than for the Lorenz model (Weiss and Brock, 1986) (see Fig. 5.2.9). It appears, therefore, that dynamics can be investigated with these systems in the homogeneous (two-level (Lorenz)) limit as well as for "coherent" (three-level) conditions.

The question as to whether the dynamical properties of the 81 µm ammonia laser resemble the Lorenz model dynamics fortuitously was investigated by experiments on

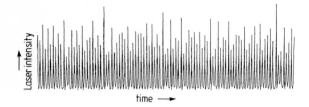

Fig. 8.2.1. "Period-doubling chaos". The pulsing period is constant, the pulse height is varying irregularly. (From Wu et al., 1989.)

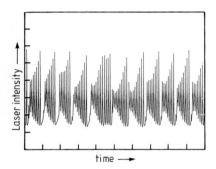

Fig. 8.2.2. Pulsing of 154 μm $^{15}NH_3$ laser as expected for Lorenz model-like behavior. (From Wu et al., 1989.)

another similar laser transition of ammonia (Wu et al., 1989), the above-mentioned 153 μm transition of $^{15}NH_3$ (Davis et al., 1981), for which similar relaxation rates are to be expected. The dynamics of this laser were found to be the same as for the 81 μm laser. In particular, the typical spiral pulses with jumps between them (Fig. 8.2.2) corresponding to the jumps between the two leaves of the Lorenz attractor are observed, together with the period-doubling transition to chaos when the laser resonator tuning is used as a control parameter.

By recording the pulsing after the pump strength is increased beyond the laser instability threshold, the destabilisation of the fixed points of the Lorenz model developing to Lorenz-chaotic pulsing is observed, as expected (Fig. 8.2.3).

A further ammonia laser transition at 374 μm has different quantum numbers from the mentioned transitions and, as mentioned above, different relaxation rates are likely. In the experiments reported (Klische and Weiss, 1985; Hogenboom et al., 1985), only periodic pulsing was observed within the pressure range making pump coherence effects unlikely and thus homogeneous, two-level dynamics likely. Under conditions where pump coherence effects are to be expected, chaotic emission was also observed. The instability (second laser) thresholds (Fig. 8.2.4) are found, in agreement with the Lorenz equations, if a large ratio of polarization to population relaxation rates ($\gamma_\perp/\gamma_\parallel > 6$) is assumed.

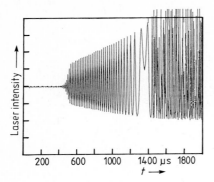

Fig. 8.2.3. Transient destabilization of the steady-state emission of the 153 μm ^{15}NH$_3$ laser. Over the length of recording, the pump strength is increased by ca. 5%. About 500 μs after the start the pump exceeds the instability threshold leading to spiraling away from the steady state. At ca. 1500 μs chaotic pulsing is reached. Fig. 8.2.2 is a magnification of the range 1600—1800 μs. The apparent low pulse frequency of the transient spiral is an artefact from interference between the digitization rate of the recording digitizer and the pulse frequency. (From Wu et al., 1989.)

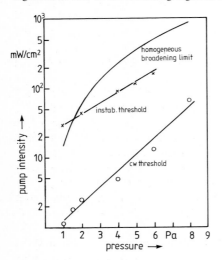

Fig. 8.2.4. Continuous laser emission threshold and threshold for instability as a function of working pressure for the 374 μm ^{15}NH$_3$ laser. (From Hogenboom et al., 1985.)

This condition appears to be likely to be met in the experiment on the grounds of the particular rotational quantum numbers ($J = 2$, $K = 0$) of the laser transition. For this relaxation rate ratio, the pulsing is predicted to be periodic and the (periodic) attractor to be symmetric (Narducci et al., 1985). The latter was confirmed by measuring the optical spectrum of the laser output under pulsing conditions; its missing central component shows that the attractor is symmetric (Fig. 8.2.5).

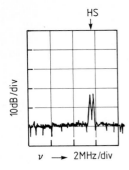

HS

10dB/div

$\nu \longrightarrow$ 2MHz/div

Fig. 8.2.5. Optical spectrum of self-pulsing $^{15}NH_3$ laser at 374 µm. The bichromatic emission shows a symmetric attractor. The Pump Rabi frequency is smaller than the homogeneous linewidth. HS: heteroclyne spectrum. (From Hogenboom et al., 1985.)

b. Theoretical Analysis

The similarity of the experimental results on optically pumped $^{14}NH_3$ and $^{15}NH_3$ lasers, in the high-pressure domain, to the predictions of the Lorenz model is not easy to understand *a priori* from a theoretical point of view. In effect, an estimate of the molecular parameters seems to indicate that, independent of the gas pressure, the relaxation rates for the pump-transition and Raman coherences are of the same order of magnitude as those of the laser-transition coherence, i.e. $\gamma_{02}, \gamma_{12} \approx \gamma_{01}$ (Pujol et al., 1988; Lawandy and Ryan, 1987). This means that condition (e) in Section 8.1, for a coherently pumped laser to be equivalent to a two-level Lorenz-model laser, is not fulfilled or, in other words, that coherent pump effects can in principle be as important as intrinsic two-level coherence effects and could lead to a non-Lorenzian behavior.

In order to investigate this problem, Pujol and co-workers (Pujol et al., 1988a and b; Laguarta et al., 1988 and 1989; Corbalán et al., 1989) numerically solved eqns. (8.1.2) for molecular and cavity relaxation rates approximately corresponding to the 81 µm $^{14}NH_3$ laser. With homogeneous broadening considered (Pujol et al., 1988a and b), the results obtained have in fact been different from the predictions of the Lorenz model, confirming that an important contribution of coherent pumping effects is present. Figure 8.2.6 shows the chaotic attractor for certain parameter values (Pujol et al., 1988a). The projection on the (field amplitude, polarization) plane, i.e. the projection on the $(\alpha/\gamma_\perp, \mathrm{Im}\varrho_{01})$ plane in Fig. 8.2.6.a, is in effect very different from the same projection in the case of the Lorenz chaotic attractor (see, for instance, Fig. 2.1.4.b), and so is the time dependence of the laser field amplitude (compare Fig. 8.2.6.a with Fig. 2.1.4.c). Further, the ratio of the second to the first laser threshold and the bifurcation diagram obtained when the pump amplitude β is continuously increased (Pujol et al., 1988a) are also very different from those corresponding to the Lorenz model. Fig. 8.2.6.b shows that the Raman coherence ϱ_{12} ruling the coherent pumping effects indeed reaches large values. Recent results obtained for slightly different values of $\kappa/\gamma_\perp$ and smaller values of the gain parameter G (the values of these parameters are not known very reliably since they depend on other poorly known molecular parameters and experimental conditions (Laguarta et al., 1988; Corbalán et al., 1989) show attractors relatively more similar to the Lorenz case (Laguarta et al., 1989) (Fig. 8.2.7), but the threshold ratio and the bifurcation diagrams remain very different from those predicted by the Lorenz model.

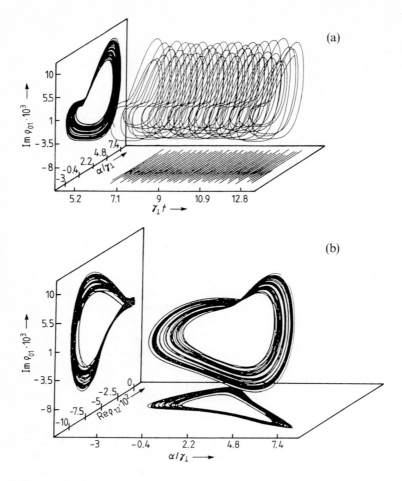

Fig. 8.2.6. Chaotic attractor corresponding to an optically pumped homogeneously broadened FIR laser for molecular and cavity relaxation rates corresponding approximately to an 81 μm $^{14}NH_3$ laser and for a pump amplitude $\beta/\gamma_\perp = 0.14$ ($\gamma_\perp \equiv \gamma_{01}$). (a) Attractor's projection on the ($\gamma_\perp t$, $\alpha/\gamma_\perp$, $Im\varrho_{01}$) space; (b) the same on the ($\alpha/\gamma_\perp$, $Re\varrho_{12}$, $Im\varrho_{01}$) space. The fixed-point solution corresponding to stationary emission lies inside the hole of the chaotic attractor. (From Pujol et al., 1988a.)

This discrepancy between experimental and theoretical results indicates that either the numerical values assigned to the different parameters are not suitable or, more fundamentally (and more probably), that some physical factors that play an important role in the laser dynamics are not included in the theoretical model yielding eqns. (8.1.2). These factors are probably those mentioned in Section 8.1: Doppler broadening (and velocity-changing collisions), M-level degeneracy, rotational manifold population reservoir and longitudinal and transverse spatial effects.

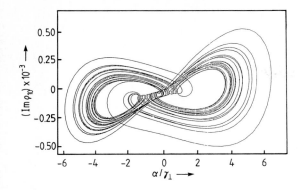

Fig. 8.2.7. Chaotic attractor obtained under conditions similar to those in Fig. 8.2.6, but for a much smaller gain parameter G and slightly different cavity losses. The attractor's projection is on the laser field $(\alpha/\gamma_\perp)$ − polarization $(\mathrm{Im}\,\varrho_{10})$ plane. (From Laguarta et al., 1989.)

Laguarta and co-workers (Laguarta et al., 1988 and 1989, Corbalán et al., 1989) specifically investigated the influence of Doppler broadening in the previous case of the 81 μm $^{14}NH_3$ laser emission. They found that this influence is very strong owing to the high sensitivity of the resonance conditions of the pump and laser fields on the molecular velocity. In fact, in the ammonia far-infrared laser the relatively small cavity losses and the large value of the electric dipole matrix element μ_{01} of the rotational laser transition allow the laser-field Rabi frequency 2α to reach values up to ca. $10\gamma_\perp$, which induce AC Stark splittings of the common level 0 (Fig. 8.1.1), which in turn strongly influence the resonance conditions of the pump field (the pump field itself also AC Stark splits the level 0, but its contribution is small under the actual operating conditions of this laser because β remains smaller than $\gamma_\perp$). As these resonance conditions are velocity dependent (through Doppler effects) and, under unstable emission conditions, the laser-field Rabi frequency 2α is time dependent, we get the result that different molecular velocity groups participate in the pumping process at different times.

In order to take into account this factor theoretically one has to solve the equations for $\dot{\varrho}_{ii}$ and $\dot{\varrho}_{ij}$ $(i,j = 0,1,2)$ of eqns. (8.1.2) for each molecular velocity group and add in the last term of the equation for $\dot{\alpha}$ all contributions to the gain (i.e. to $\mathrm{Im}\,\varrho_{01}^*$) of these groups. One also has to add in the last term of eqn. (8.1.3) the contributions to the dispersion (i.e. to $\mathrm{Re}\,\varrho_{01}$) of all the velocity groups. Figures 8.2.8 and 8.2.9 show the results obtained in this way (Corbalán et al., 1989; Laguarta et al., 1989; Roldán et al., 1989). Figure 8.2.8 corresponds to central tuning (i.e. to a cavity tuning yielding maximum gain), for which chaotic behavior appears, whereas Fig. 8.2.9 corresponds to other tunings, for which periodic behavior is obtained.

The principal result is that the bifurcation diagrams and the laser intensity look qualitatively similar to those predicted by the Lorenz model (compare Fig. 8.2.8.a with the analysis in Section 2.1, and Fig. 8.2.8.c with Fig. 2.1.4.c) or by the extended Lorenz model in Section 5.2.1.c (compare Fig. 8.2.9.a with Fig. 5.2.1 for a fixed value of R). In particular, one finds a sudden appearance of a chaotic attractor followed by an inverse sequence of period doublings when the pump strength is increased (Fig. 8.2.8.a), a

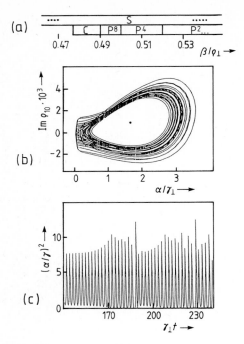

(a)

$\beta/\varrho_\perp \rightarrow$

(b)

$\alpha/\gamma_\perp \rightarrow$

(c)

$\gamma_\perp t \rightarrow$

Fig. 8.2.8. Results obtained by taking into account Doppler broadening, for conditions corresponding approximately to an 81 μm ^{14}NH$_3$ laser operating at 9 Pa with 30 MHz off-center pumping and tuned to maximum gain ($\Delta_1^c = -5\gamma_\perp$). (a) Bifurcation diagram obtained by continuously increasing the pump amplitude β; the upper strip corresponds to the fixed-point solution yielding steady-state emission, the lower strip to the large attractor (S = stable, C = chaotic, P^n = periodic behavior of period n-times the fundamental period). (b) Projection of the chaotic attractor on the (laser field amplitude, gain) plane for $\beta/\gamma_\perp = 0.478$; the dot denotes the fixed-point solution. (c) Time dependence of the laser intensity for the chaotic attractor in (b). (From Corbalán et al., 1989.)

complete Feigenbaum scenario when the laser cavity is tuned towards maximum gain (Fig. 8.2.9.a) and an adequate second-to-first threshold ratio (ca. 14). Hence one may conclude that inhomogeneous broadening, i.e. the contribution of different molecular velocity groups, results in the apparent disappearance of coherent pumping effects in the observed dynamic behavior; surprisingly, features typical of the two-level homogeneously broadened laser are observed. This provides a basis for the understanding of the experimentally observed behavior of the ammonia laser.

Apparently some differences from the Lorenz model remain, however. One of them is the attractor's symmetry. Figure 8.2.10 shows a comparison between the attractor of Fig. 8.2.8.b and the Lorenz attractor. In the first case the attractor is one-sided. The opposite attractor, which also exists, is also shown (note that the system of eqns. (8.1.2) is symmetric with respect to the sign of the variables ϱ_{01}, ϱ_{12} and α); however, the two one-sided attractors are not dynamically coupled, i.e. there is a small separation be-

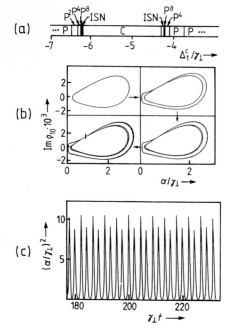

(a)

(b)

(c)

Fig. 8.2.9. The same as in Fig. 8.2.8, showing (a) bifurcation diagram of the large attractor as a function of the cavity detuning Δ_1^c, for $\beta/\gamma_\perp = 0.478$ (ISN = inverse sequence of noisy periods); the maximum gain corresponds to the center of the bifurcation diagram ($\Delta_1^c/\gamma_\perp \approx -5$); (b) attractor's projection for $\Delta_1^c/\gamma_\perp = -3, -4, -4.2$ and -4.23 in the sense indicated by arrows; (c) time plot of the laser intensity in the P^8 regime at $\Delta_1^c/\gamma_\perp = -4.23$. (From Corbalán et al., 1989.)

tween them which can only be overcome by the influence of noise. Therefore, in the case of the optically pumped laser (Fig. 8.2.10.a), the attractor seems to be asymmetric and the sign of the laser field amplitude α remains either positive or negative with evolution of time, whereas in the Lorenz model (b) the attractor is symmetric and the field alternates in sign after each spiraling.

A recent analysis (Roldán et al., 1989), however, reveals that the asymmetry of the attractor in the first case is only apparent. If one takes into account that the laser field amplitude is complex, i.e. it is proportional to $\alpha \exp(-i\phi)$, and one calculates the time evolution of the phase ϕ, one finds that each time the amplitude α comes close to zero the phase ϕ undergoes a sudden change of approximately π rad (Fig. 8.2.11). Hence the field amplitude that can be detected, for instance by a heterodyne detection technique, will actually show changes of sign as in the case of the Lorenz attractor.

Recent experimental investigations on the 153 μm $^{15}NH_3$ laser (Weiss et al., 1988) have revealed the existence of both kinds of attractors: an asymmetric one for low gas pressures and a symmetric or Lorenz-type one for high pressures (Fig. 8.2.12), which support "homoclinic" and "heteroclinic" chaos, respectively. The former remains to

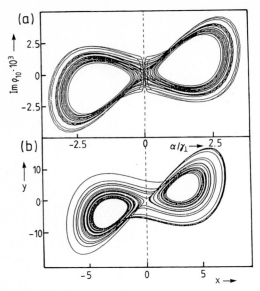

Fig. 8.2.10. (a) The chaotic attractor in Fig. 8.2.8b together with its opposite attractor (see text). (b) The Lorenz attractor for physical conditions as close as possible to those considered in Fig. 8.2.8. (From Corbalán et al., 1989.)

be explained theoretically; presumably, the influence of optical pumping effects is larger at low pressures and is not fully counterbalanced by Doppler broadening.

In the case of the 374 μm $^{15}NH_3$ laser emission, which, as mentioned in Section a, shows periodic instead of chaotic behavior, the interpretation made by Klische and co-workers (Klische and Weiss, 1985; Hogenboom et al., 1985) in terms of the Lorenz model has been disputed by others (Lawandy and Plant, 1986; Lawandy and Ryan, 1987). They argued that the condition $\gamma_\parallel/\gamma_\perp < 0.2$ predicted by the Lorenz model for observing periodic behavior (Narducci et al., 1985; Sparrow, 1982) cannot be met for such a laser transition (they calculated a value $\gamma_\parallel/\gamma_\perp \approx 1$), and that coherent pumping effects should be the actual cause of the observed self-pulsing. In this respect, Khandokhin et al. (1988) solved equations similar to eqns. (8.1.2) under resonant conditions ($\Delta_2 = \Delta_1^c = \Delta_1 = 0$) and with a simplified treatment of the molecular relaxation, and found a qualitative agreement between their results and the experiment of Hogenboom et al. (1985) under conditions where coherent pumping effects are apparent and the Lorenz model cannot describe the system (see Fig. 8.2.13). This agreement, however, occurs only for $\gamma_\parallel/\gamma_\perp \approx 0.4$ and not for $\gamma_\parallel/\gamma_\perp = 1$, which would lead to an unrealistically high second threshold. The various different findings can only be clarified when the relaxation rates of the laser levels become known.

Ryan and Lawandy (1987) also solved equations similar to eqns. (8.1.2) under resonant conditions and in the limit $\gamma_{02}, \gamma_{12} \gg \gamma_{ii}, \gamma_{01}, \alpha$ ($i = 0, 1, 2$) for which pump and Raman coherences are absent. This means that the only effect of optical pumping is the change in pump absorption, brought about by the time dependence of the upper

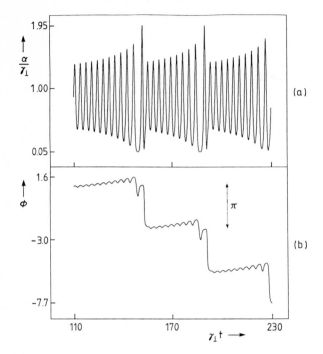

Fig. 8.2.11. Temporal evolution of the field amplitude (a) and of the phase (b), for a case of chaotic behavior similar to that in Fig. 8.2.8 (from Roldán et al., 1989).

level population. They found, for $\gamma_\parallel/\gamma_\perp \approx 1$, a very low instability threshold (1.6 times above the laser threshold) at which periodic pulsing appears. By increasing the pump intensity a period-doubling transition to chaos is obtained. In fact, a qualitatively similar behavior is observed experimentally on the 81 and 153 μm ammonia lasers, in the low-pressure domain. At low pressures (see Fig. 5.2.9), the measured laser instability threshold is much lower than that predicted from the Lorenz model. Experimentally, the onset of chaos under these conditions is found to occur via period doublings when the pump strength is increased (Weiss, unpublished work).

A different theoretical approach was followed by Harrison and co-workers (Mehendale et al., 1986a and b; Moloney et al., 1987; Uppal et al., 1987), who performed a detailed analysis of the dynamic behavior in the case of homogeneous broadening for several parameter sets without attempting a direct comparison with experimental results. By solving eqns. (8.1.2) for steady-state conditions, they calculated the complex susceptibility associated with the laser transition, i.e. they determined the gain and dispersion line-shapes as a function of laser field detuning. They showed that for low laser intensities and by increasing the pump field strength beyond saturation, the gain (and the laser emission profile) splits into two peaks (as a result of pump-induced AC Stark splitting) and, simultaneously, the dispersion curve undergoes large changes

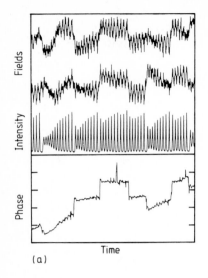

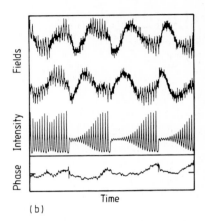

Fig. 8.2.12. (a) High-pressure (9 Pa) chaotic pulsing of 153 μm ^{15}NH$_3$ laser emission for resonant tuning. Trace marked "intensity": laser intensity pulses. Pulsing period, 1 μs. Traces marked "fields": in-phase and in-quadrature heterodyne signals measuring the laser field. Trace marked "phase": phase changes of the laser field as a function of time, reconstructed from field traces. One division on the vertical axis corresponds to a phase change of π rad. The phase change of π at the end of each spiral indicates a change of field sign (symmetric attractor). (From Weiss et al., 1988.) (b) Chaotic pulsing of 153 μm ^{15}NH$_3$ laser emission for resonant tuning and a pressure of 5 Pa. Traces as (a) with same time and phase scales. Note the absence of π phase jumps, which indicates always the same sign of the field (asymmetric attractor). (From Weiss et al., 1988.)

(anomalous dispersion shape at the line center) which can compensate for the wavelength shift brought about by gain splitting (Fig. 8.2.14).

If this occurs, a new kind of "spontaneous mode splitting" (or "pump-induced mode splitting") may appear, i.e. two frequencies corresponding to a single resonator mode can be simultaneously amplified and regular undamped oscillations may appear (Mehendale and Harrison (1986a and b). This mode splitting is different from those found in homogeneously (Section 5.2.1) and inhomogeneously (Section 5.3.1) broadened two-level lasers.Up to now the restricted conditions under which it could be observed have not been encountered in experimental optically pumped lasers.

The same group also performed bifurcation analyses in some regions of the control parameter space. They found Hopf bifurcations, mode splitting, periodic pulsations, chaos (Moloney et al., 1987) (Fig. 8.2.15) and a codimension-2 point for which homoclinic orbits appear (Forysiak et al., 1989).

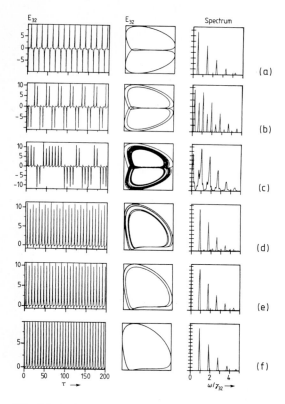

Fig. 8.2.13. Field pulses, attractor projections in the field-inversion plane and spectra calculated for fixed parameters, except for the homogeneous linewidth, which is varied. On varying the homogeneous linewidth ((a) to (c)), a symmetric attractor period doubles to chaos. From the other side ((f) to (c)) an asymmetric attractor period doubles to the same chaos. Note that again chaos occurs where two different oscillation forms compete. The uppermost case is assumed to correspond to the measurements in Fig. 8.2.5. (From Khandokhin et al., 1988.)

8.2.2 Standing-Wave Lasers

Lefebvre et al. (1984) performed experiments on a Fabry-Pérot formic acid laser optically pumped by a single-frequency CO_2 laser in which the pump beam enters the laser cavity through one mirror and is reflected back by the second mirror, so that pumping is bi-directional. Spontaneous pulsing in the single-mode emission at 742 μm was observed when the cavity was tuned to resonance (i.e. to the center of the laser transition) and the CO_2 laser of high intensity was sufficiently detuned from the resonance of the pump transition to excite two distinct velocity groups of molecules contributing to the laser gain.

Mandel and co-workers (Abraham et al., 1985; Wu and Mandel, 1985) presented a simple theoretical model that provides a basis for understanding the origin of these

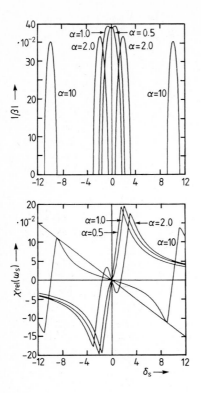

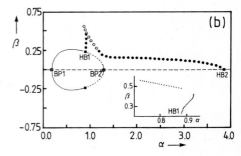

Fig. 8.2.14. Normalized laser amplitude β and dispersion $\chi'_{rel}(\omega_s)$ versus normalized detuning δ_s for different pump amplitudes α (with the notation of Fig. 8.1.1 and eqns. (8.1.2), β, α, ω_s and δ_s correspond to $\alpha/\gamma_\perp$, $\beta/\gamma_\perp$, ω_L and $\Delta_1/\gamma_\perp$, respectively, where $\gamma_\perp \equiv \gamma_{01}$). Also plotted in the lower figure is the straight line respresenting the cavity resonance condition. The intersection of this line with each plotted curve determines the frequencies which are resonant with the cavity. Other parameter values are $\gamma_0 = \gamma_1 = \gamma_2 = \gamma_\parallel = 0.2\,\gamma_\perp$, $\gamma_{01} = \gamma_{02} = \gamma_{12} = \gamma_\perp$, $\sigma = \kappa/\gamma_\perp = 10$, $G/\gamma_\perp^2 = 80$. Note that for $1.5 < \alpha < 10$ there is mode splitting, i.e. there is a pair of symmetric frequencies simultaneously possessing net gain and resonance conditions. (From Uppal et al., 1987.)

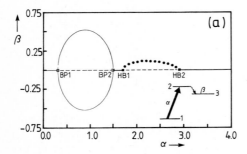

Fig. 8.2.15. Bifurcation diagrams for $G/\gamma_\perp^2 = 50$, $\sigma \equiv \kappa/\gamma_\perp = 10$, $b \equiv \gamma_\perp/\gamma_\parallel = 0.8$ (a) and b = 0.2 (b). The plots show the laser-emission amplitude β ($\alpha/\gamma_\perp$ in our notation) at the line center versus pump amplitude α ($\beta/\gamma_\perp$ in our notation). The solid lines denote stable cw operation and the dashed lines unstable cw operation; the filled circles are branches of stable periodic solutions and the open circles branches of unstable periodic solutions. Inset in (a): schematic diagram of the three-level system under study. Inset in (b): blow-up of the region near HB1. HB1 and HB2 are Hopf bifurcation points and BP1 is a pitchfork-bifurcation point. (From Moloney et al., 1987.)

instabilities. In this model, the laser medium is formed by two distinct resonant species, i.e. by two homogeneously broadened groups of two-level atoms with different resonant frequencies. These groups of atoms represent the molecular velocity groups that are excited by the two counter-propagating pump beams, whose resonant frequencies are Doppler shifted. The unsaturated gain is then the sum of two Lorentzian lines (Fig. 8.2.16.a) and the unsaturated dispersion behaves as shown in Fig. 8.2.16.b.

The gain and dispersion line shapes shown in Fig. 8.2.16 display qualitative similarities to those in Fig. 5.3.6. Hence they can give rise to a new kind of "spontaneous mode splitting" (or mode splitting induced by the pump field) different from that mentioned in Chapter 5 and that in Fig. 8.2.14. When the laser cavity is tuned to resonance, two symmetric frequencies having positive gain (i.e. frequencies close to the gain peaks in Fig. 8.2.16.a) are simultaneously resonant owing to the anomalous dispersion shown in Fig. 8.2.16.b. In this way the experimentally observed self-pulsing can be explained as the simultaneous amplification of both frequencies. Other predictions of the model, such as stability analysis, phase diagrams and time evolutions, have been described (Abraham et al., 1985; Wu and Mandel, 1985).

This model has notable limitations, namely that it neglects all coherent pumping effects, and ignores the standing-wave nature of the laser field. However, for resonant tuning of the laser cavity and moderate pump field strengths the model explains several of the observed features. From another point of view, this model can be considered as a two-level laser model with the simplest conceivable kind of inhomogeneous broadening, because one has only two resonant atomic frequencies instead of, as is usual, a continuous distribution of resonant atomic frequencies (in this sense, it could also be included in Section 5.3).

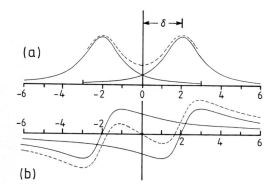

Fig. 8.2.16. (a) Unsaturated gain and (b) dispersion for two resonant groups of atoms with their resonant frequencies separated by a detuning parameter value of $\delta = 2$. The solid curves show contributions of the two groups separately, and the dashed curves show the total gain and dispersion. (From Abraham et al., 1985.)

8.3 Other Cases

8.3.1 Cw Mid-Infrared Lasers

Although the most detailed experiments have been carried out using ammonia far-infrared lasers, experiments have also shown chaotic dynamics of single-mode optically pumped lasers operating on vibrational and electronic transitions in the mid-infrared and the near-visible ranges. In the mid-infrared region ammonia was again used as the active medium. In contrast to the far-infrared lasers, in this case population inversion and lasing are created on a vibrational transition (see Fig. 8.3.1) (Siemsen et al., 1987). Figure 8.3.2 shows the optical arrangement used for observing the dynamics of a 12 μm ammonia laser. The pump laser operates on the $9R(30)$ CO_2 line. Its frequency is shifted by 180 MHz using two acoustooptic frequency shifters to tune the pump laser radiation into resonance with the $sR(5,0)$ $^{14}NH_3$ vibrational transition. A $^{14}NH_3$ low-pressure gas cell is utilized to find the $sR(5,0)$ transition line center using a Lamb-dip technique. The pump laser radiation is then coupled into the 12 μm ammonia laser. A grating is

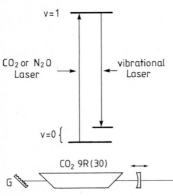

Fig. 8.3.1. Energy level scheme of laser-pumped mid-infrared gas lasers. In contrast to the far-infrared laser, the lower laser level lies in the vibrational ground state.

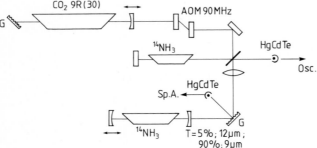

Fig. 8.3.2. Set-up for the measurement of pulsing instabilities of mid-infrared laser-pumped gas lasers. The pump source is a CO_2 laser tuned by a grating. Its frequency is shifted into resonance with an absorption line of ammonia by an acousto-optic modulator (AOM). The $^{14}NH_3$ cell serves as an absorption line-center indicator by a Lamb-dip technique. The laser under investigation receives its pump radiation via a grating (G) which permits the pump radiation to be spatially separated from the generated 12 μm radiation; T: transmission. (From Siemsen et al., 1987.)

used to separate the pump radiation at 9.3 μm from the generated 12 μm radiation. The output of the fast HgCdTe photodetector is fed to an RF spectrum analyzer to obtain the intensity spectrum of the laser output.

Figure 8.3.3 shows a series of intensity spectra thus observed when the 12 μm laser resonator was tuned towards the line center. The continuous noise spectrum observed at the line center is preceded by periodic pulsing and two subharmonics, thus giving a clear indication of a transition to chaos via period doubling. Although this is again qualitatively the dynamics predicted by Ryan and Lawandy (1987), it should be noted that in the experiment, in contrast to the model, a standing-wave laser was used, which has the complication of a spatially varying field in the medium. At the gas pressure used, the "bad-cavity" condition is fulfilled.

8.3.2 Cw Near-Infrared and Visible Lasers

An optically pumped gas laser emitting on an electronic transition near the visible range was tested (Wu and Weiss, 1987). In order to fulfil easily the "bad-cavity" condition with a low-loss resonator, iodine was chosen as the active gas. Its homo-

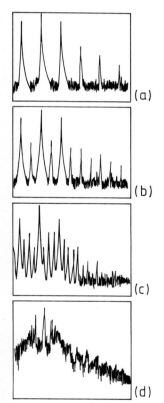

Fig. 8.3.3. Intensity spectra of mid-infrared self-pulsing gas laser. Spectra are taken at different detunings. A clear period-doubling sequence to chaos when tuning towards the gain line center (from (a) to (d)) can be recognized. (From Siemsen et al., 1987.)

geneous linewidth at a few 100 mTorr is a few hundred kHz. Thus a resonator with a few percent loss fulfils the "bad-cavity" condition. The iodine pressure in the gas cells was controlled by the temperature using Peltier-cooled cold-fingers on the side of the cells.

To make the system more readily comparable to models, a ring resonator was chosen as shown in Fig. 8.3.4. The strongest iodine laser transitions (highest gain) at 1340 nm were chosen so as to reach more easily a second laser threshold.

The iodine-dimer laser was optically pumped by a single-frequency Ar^+ laser. The emission linewidth of the latter was reduced to below the homogeneous iodine linewidth by a fast frequency control which uses the slope of a transmission Fabry-Pérot interferometer as a frequency discriminator. The Fabry-Pérot had a length of 10 cm and a finesse of 100, so that the linewidth of the Fabry-Pérot was ca. 7 MHz. With a control bandwith of 3 kHz, the laser linewidth was reduced to 100 kHz. An iodine cell was again used as an absolute pump frequency marker using the Lamb dips of the various iodine hyperfine lines falling within the tuning range of the 514 nm Ar^+ laser line.

Since the 514 nm Ar^+ laser, irrespective of its frequency setting, pumps many hyperfine transitions at ca. 1340 nm, conditions for single line emission of the laser could only be found at certain pump frequency and iodine laser resonator setting combinations. Uni-directional (backward wave) laser emission was achieved in the same way as for the far-infrared ammonia lasers utilizing the Doppler effects. Under single frequency — single line — single direction emission conditions, pulsing instabilities could be observed also for this laser.

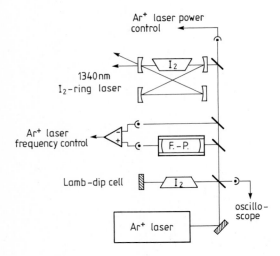

Fig. 8.3.4. Set-up for the measurement of pulsing instabilities of an electronic transition, laser-pumped iodine gas laser at 1.3 μm. The laser under investigation is a ring laser. The single-mode 514 nm Ar^+ laser is frequency controlled to a Fabry-Pérot (F.-P.) resonator to reduce its linewidth below the homogeneous linewidth of the active iodine gas. The further Lamb-dip iodine cell permits the pump laser frequency to be set reproducibly. (From Wu and Weiss, 1987.)

Figure 8.3.5 shows a series of intensity spectra obtained when tuning the iodine laser resonator progressively closer to the gain line center, all other system parameters being held constant. The familiar period-doubling transition to chaos is observed (Wu and Weiss, 1987).

It is likely that the instabilities of the mid-infrared and near-visible laser investigated are not the homogeneously broadened laser instabilities that correspond to the Lorenz model; rather, they probably involve three-level coherences.

In any case, these experiments show that for the large class of laser-pumped gas lasers that provide a few thousand laser transitions from the millimeter wave range to the visible spectral range, instabilities and even chaotic emission are typical and not the exception, as had been believed before.

The initial experiments open up a large field for investigations of laser instabilities. The class of lasers offers a wide range of parameters and basically different laser properties which should allow various laser models to be tested. More detailed information on the dynamics of these lasers can easily be obtained by observations of pulse shapes and heterodyne (laser field) measurements, which are particularly simple to do in the short-wavelength range. In particular, the electronic transition lasers allow the fluorescence from the upper laser level to be measured simultaneously with the laser intensity, thus giving a measure of the time-dependent population inversion. Detailed comparisons with calculations and more reliable attractor reconstructions will therefore possible. The attainment of the conditions for the Lorenz model is also particularly easy since modulators are available for visible radiation that permit phase noise modulation of the pump to make the pump effectively incoherent.

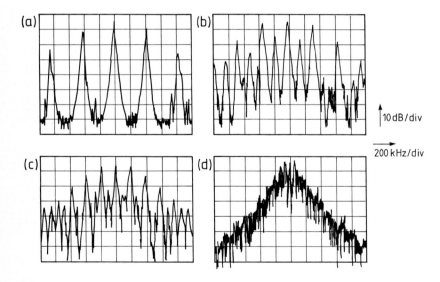

10 dB/div

200 kHz/div

Fig. 8.3.5. A period-doubling transition to chaos measured on the iodine laser when tuning the iodine laser resonator towards the gain line center (from (a) to (d)). Intensity spectra. (From Wu and Weiss, 1987.)

Since experiments are technically most easily done in the visible spectral range, an experiment on instabilities of a 525 nm sodium-dimer laser has also been performed (Weiss and Klische, unpublished work). This laser appeared to be preferable to, e.g., the visible laser transitions of iodine since it shows only a few transitions. Hence the complication of the multitude of (hyperfine) laser transitions is avoided. On the other hand, the homogeneous linewidth of the Na_2 transitions is approximately a factor of 500 larger than for I_2. This implies that, to attain "bad-cavity" instabilities, a gain also 500 (!) times larger than for the 1340 nm iodine laser is necessary. An estimate on the basis of published gain data for sodium laser transitions (Wellegehausen et al., 1977) shows that this condition might just be in reach using an optimally short resonator length for the 525 nm laser transition which is pumped by a 488 nm (single-frequency) Ar^+ laser.

The laser resonator used is shown in Fig. 8.3.6. It is a ring resonator with an internal prism for (wavelength) selection of the 525 nm Na_2 line, also used as a convenient way of coupling the pump radiation into the laser resonator. The buffer gas used in the sodium cell is helium. At a helium pressure of 3 Torr and a sodium temperature of 550°C, self-pulsing and one period doubling are observed (Fig. 8.3.7). It is found that the appearance of the half-pulsing frequency (Fig. 8.3.7b) may be related to a transverse mode which locks, so that its frequency difference from the initial mode is half of the pulsing frequency. More precise information can only be obtained by heterodyne measurements. It should be possible, however, to suppress off-axis transverse modes and reach higher instabilities and chaos if pump powers higher than 500 mW can be applied. Owing to the visibility of the laser emission, this laser has also permitted transverse effects, i.e. possibly different dynamics in different parts of the laser mode, to be investigated (Klische et al., 1989).

8.3.3   Pulsed Mid-Infrared Lasers

When the pressure of an optically pumped ammonia laser is raised to a few Torr, the rotational relaxation becomes fast so that pump intensities in the MW/cm^2 range have

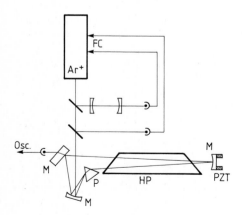

Fig. 8.3.6. Sodium-dimer ring laser pumped by a single-mode 488 nm Ar^+ laser. (From Klische et al., 1989.) FC: laser frequency control, M: mirror, P: prism, HP: Na_2-heat-pipe, PZT: piezo translator.

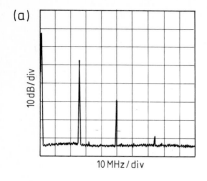
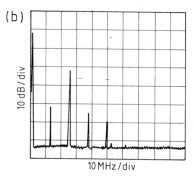

Fig. 8.3.7. (a) Self-pulsing and (b) period-doubling intensity spectra of the sodium laser. (From Weiss and Klische, unpublished work.)

to be used to achieve lasing. Obviously this is possible only for pulsed operation. An experiment was carried out to demonstrate instabilities also under these conditions (Harrison et al., 1985).

The ammonia laser working near 12.8 μm consisted of a linear resonator of 25 cm length in which a Brewster-window ammonia cell of length 22 cm was mounted. This resonator was optically pumped by a pulsed high-pressure CO_2 laser oscillating in one mode, which provided pulses of 2 μs full width at half-maximum. The separation between the pump laser line center and the ammonia absorption line center is 1.3 GHz; exceeding the tuning range of the pump CO_2 laser. Therefore, as has also been verified by experiments, the ammonia laser emission is an off-resonant Raman-type two-photon transition. One photon is absorbed simultaneously with one photon emitted, both photons being mismatched in frequency with the ammonia transitions by the same amount. The working pressure of 8 Torr is so high that all Doppler effects are masked by the homogeneous pressure broadening.

When tuning the ammonia laser resonator away from line center there is "steady-state" laser output. Closer to the line center, pulsing with a period of 18 ns is observed (Fig. 8.3.8a), even closer to the line center an apparent doubling of the pulsing period occurs (Fig. 8.3.8b), and on the line center an irregular pulsing (Fig. 8.3.8c) occurs. This sequence is qualitatively similar to what was observed with all the continuously operating optically pumped systems.

Objections have been raised as to whether the situation really corresponds to asymptotic (attractor) conditions or shows a transient phase. It was mentioned (Harrison et al., 1985) that the pumping pulse lasts for 60 pulsing cycles, on which basis it is argued that the experiment corresponds to asymptotic conditions. (Numerical calculations of the Lorenz model, however, easily show transients up to a few hundred pulses long, and this corresponds to what is seen in cw experiments (Wu et al., 1989).

It is pointed out that this laser works under homogeneous broadening transitions. The instability threshold is found about two times above the laser threshold (for "bad-cavity" conditions only). The latter contradicts the possible interpretation as Lorenz-type dynamics. In fact, one might argue that at the high pump intensities used, the AC

100ns   50ns

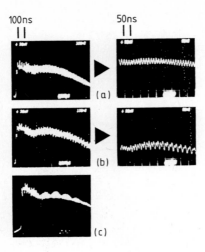

Fig. 8.3.8. Single-mode instability of a pulsed mid-infrared ammonia laser: (a) pulsing; (b) period-doubling; (c) chaos. For clarity, magnifications are shown for (a) and (b). (From Harrison and Biswas, 1985.)

Stark shifts (no splittings since pumping is strongly non-resonant) lead to different emission frequencies of the various magnetic sub-transitions that would then again produce some inhomogeneous broadening, making the relatively low instability threshold somewhat plausible.

9 Multimode Lasers

Longitudinal multimode emission in a laser occurs usually when the amplifying medium is inhomogeneously broadened and the inhomogeneous width of the lasing transition is larger than the intermode frequency interval $c/2L$ (where c represents the velocity of light in the medium and $2L$ the resonator round-trip length), because each mode is supported by a different group of atoms. Multimode emission also occurs for standing-wave resonators owing to spatial hole burning (see any text on laser physics, e.g. Haken (1985)). In these cases, the different modes are relatively independent of each other.

However, longitudinal multimode emission can also appear under simpler conditions (for instance, in homogeneously broadened ring-resonator lasers), and may give rise to interesting dynamical processes such as passive mode-locking and various kinds of instabilities which we report in this chapter. We consider in Section 9.1 the case of co-propagating longitudinal modes and in Section 9.2 the case of counter-propagating modes of equal frequency. Finally, in Section 9.3 other special cases such as lasers with intracavity frequency doubling, transverse modes, diode lasers with external reflector and instabilities in active mode-locking are discussed. The general concepts of active and passive mode locking are described elsewhere (Haken, 1985).

9.1   Co-Propagating Longitudinal Modes

In this section we consider cases where the simultaneously emitting modes are co-propagating modes of a ring-cavity laser with a homogeneously broadened medium.

9.1.1   Homogeneously Broadened Two-Level Lasers

In Section 5.2 we considered the stability of a homogeneously broadened single-mode uni-directional laser, and found that under "bad-cavity" conditions (eq. (5.2.8)) a "second laser threshold" exists, which is given by eq. (5.2.7), at which the single-mode emission becomes unstable. There the possibility of excitation of further cavity modes other than that considered was not taken into account, which can be correct when the mode separation is large enough.

When considering several longitudinal modes, however, a different "second laser threshold" is found where the single-mode laser emission turns into multimode unstable emission. The conditions for the appearance of this kind of instability are different from those of the single-mode instability.

The multimode instability was predicted theoretically by Risken and Nummedal (1968a and b) and Graham and Haken (1968) and its physical origin can be explained in the following way (Hendow and Sargent III, 1982; Hillman et al., 1982; Sargent III et al., 1983). Let us first recall that in an incoherent picture the homogeneously broadened laser would be single mode and stable, irrespective of how many resonator modes lie within the gain line profile: a mode closest to the line center will saturate the gain line uniformly at the mode frequency, down to the laser threshold, without distorting its Lorentzian line shape. Other modes lying further away from the line center are then by definition below threshold. However, taking into account the coherence of the interaction between the laser field and the medium, Rabi sidebands are created (Fig. 9.1.1), as in the case of the homogeneously broadened single-mode laser. At the position of the Rabi sidebands the gain may then be higher than the laser threshold gain. If by a proper choice of the laser resonator length, longitudinal modes are placed at the frequencies of the Rabi sidebands, these will be above threshold. Evidently, owing to the interaction of the modes above threshold, these may lock, producing regular laser pulsing, or may interact so that the laser pulsing becomes chaotic.

Clearly, the conditions for encountering this laser instability are less restrictive than for the single-mode laser instability because for the latter proper gain *and* dispersion must be furnished by the active medium ("mode splitting"; see Chapter 5), whereas for multimode instability only the proper gain conditions have to be provided by the medium. The dispersion is taken care of largely by the resonator. Thus, in contrast to the single-mode laser instability, it furthers the instability if the dispersion of the laser resonator dominates that of the medium. Hence a "good" cavity is favorable as opposed to the strict "bad-cavity" requirement of the single-mode laser instability. The tran-

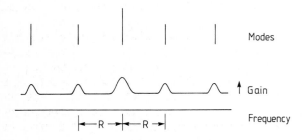

Fig. 9.1.1. Scheme of multimode laser instability. A strong field at one longitudinal mode generates Rabi sidebands. If the Rabi frequency is equal to the free spectral range, lasing can occur at other longitudinal mode frequencies; if the various modes are synchronized, periodic pulsing occurs.

sition Rabi frequencies that must be achieved and with them the degree of pumping above laser threshold are similar to the case of the single-mode instability. However, the excess factor by which the laser has to be pumped over the first threshold refers in the multimode case to a low threshold ("good cavity") and thus in absolute pump strength is considerably lower than for the single-mode instability. It is therefore surprising that no experiment is known in which multimode instabilities could be clearly attributed to the mechanism described (see in this respect Section 9.1.2). It is safe to say that in all systems for which single-mode instabilities have been observed, this multimode instability should also occur.

To explain and describe theoretically this basic kind of multimode instability, one has to generalize the homogeneously broadened uni-directional laser equations (5.1.19) slightly in order to allow for multimode emission (Haken, 1985). Indeed, if several modes are amplified, a spatial dependence of the total field amplitude will appear as a result of mode beating, i.e. $E(t)$ transforms into $E(x,t)$, where x represents the position along the cavity axis. As a consequence, the coupled variables $p(t)$ and $d(t)$ also become space dependent: $p(x,t)$ and $d(x,t)$, respectively. In this case, the general wave equation of the electromagnetic theory indicates that the transformation (Haken, 1985)

$$\partial/\partial t \rightarrow \partial/\partial t + c\,\partial/\partial x \tag{9.1.1}$$

has to be introduced into the first equation of (5.1.19). If, for simplicity, we assume that one of the cavity modes (namely the mode that is lasing below the second threshold) is in resonance with the atomic frequency, i.e. $\omega_c = \omega_{ab} = \omega$, and further we omit from $E(x,t)$ and $p(x,t)$ the rapidly oscillating factor $\exp(-i\omega t)$:

$$E(x,t) = \tilde{E}(x,t)e^{-i\omega t}$$
$$p(x,t) = \tilde{p}(x,t)e^{-i\omega t} \tag{9.1.2}$$
$$d(x,t) = \tilde{d}(x,t),$$

and we normalize the amplitudes $\tilde{E}(x,t)$, $\tilde{p}(x,t)$ and $\tilde{d}(x,t)$ to their stationary values $\bar{E}$, $\bar{p}$ and $\bar{d}$, respectively (which describe the cw laser emission), eqns. (5.1.19) transform into the standard set (Graham and Haken, 1968; Risken and Nummedal 1968a and b; Haken, 1985):

$$\{\partial/\partial t + c\,\partial/\partial x\}\hat{E} = -\kappa(\hat{E} - \hat{P})$$
$$\partial/\partial t\,\hat{P} = -\gamma_\perp(\hat{P} - \hat{E}\hat{D}) \tag{9.1.3}$$
$$\partial/\partial t\,\hat{D} = -\gamma_\parallel[\hat{D} - 1 - \Lambda + \Lambda(\hat{E}^*\hat{P} + \hat{E}\hat{P}^*)/2]$$

where

$$\hat{E}(x,t) = \tilde{E}(x,t)/\bar{E}$$
$$\hat{P}(x,t) = \tilde{p}(x,t)/\bar{P} \tag{9.1.4}$$
$$\hat{D}(x,t) = \tilde{d}(x,t)/\bar{d}$$

and

$$\Lambda = (\bar{d}_0 - \bar{d})/\bar{d}. \tag{9.1.5}$$

If we express the complex amplitudes $\hat{E}$ and $\hat{P}$ in the form

$$\hat{E} = |\hat{E}|e^{i\phi}, \quad \hat{P} = |\hat{P}|e^{i\psi} \tag{9.1.6}$$

then the cw solution of eqns. (9.1.3) is obviously

$$|\hat{E}|_{cw} = |\hat{P}|_{cw} = |\hat{D}|_{cw} = 1 \tag{9.1.7}$$
$$\phi = \Psi = \text{const.}$$

where without loss of generality (except for the problems of phase diffusion in cases of presence of noise; see references in Risken and Nummedal (1968a and b)) we can set the constant phase equal to zero.

In order to investigate the stability of this solution one may study the dynamical evolution of a solution close to the cw solution:

$$\hat{E} = (1 + e)e^{i\phi}; \quad \hat{P} = (1 + p)e^{i\psi}; \quad \hat{D} = 1 + d \tag{9.1.8}$$

where e, p, d, ϕ and Ψ are small quantities. We substitute eq. (9.1.8) into eqns. (9.1.3), retain only the linear terms and solve by the ansatz:

$$(e, p, d, \phi, \Psi) = (e_0, p_0, d_0, \phi_0)e^{i\alpha x/c + \beta t} + \text{c.c.} \tag{9.1.9}$$

In this way, the stability criterion is given by the sign of β: the cw solution (9.1.7) is stable if for every α one has $\text{Re}\{\beta\} \leqslant 0$, and it is unstable if for (at least) one value of α one has $\text{Re}\{\beta\} > 0$.

In the absence of boundary conditions α would be continuous. However, in a ring resonator of length L the variables have to fulfil periodic boundary conditions, so that α can only take the discrete values

$$\alpha_n = 2\pi c n/L, \quad n = 0, \pm 1, \pm 2,\ldots \tag{9.1.10}$$

Thus, the existence of solutions of eqns. (9.1.3) of the type (9.1.9) with values of β fulfilling $\text{Re}\{\beta\} > 0$ for one or several values of α_n with $n \neq 0$ means that some of the off-resonance modes $\omega \pm \alpha_n$ are unstable. In other words, the cw solution (9.1.7) loses its stability by converting to multimode emission.

Introduction of eqns. (9.1.8) and (9.1.9) into eqns. (9.1.3) in the way mentioned above confirms (Haken, 1985; Risken and Nummedal, 1968; Graham and Haken, 1968) that the multimode instability does not require (unlike the single-mode Lorenz-type instability) the "bad-cavity" condition $\kappa > \gamma_\perp + \gamma_\parallel$ (see eq. (5.2.8)), and the threshold condition is

212   *9 Multimode Lasers*

$$r > r_c = 5 + 3b + [8(b + 1)(b + 2)]^{1/2} \qquad (9.1.11)$$

where $r = \Lambda + 1$ and $b = \gamma_\parallel / \gamma_\perp$ have the same meaning as in the single-mode Lorenz-Haken model (Section 5.2). The second threshold or instability threshold (9.1.11) is different from that found in the single-mode case (see eq. (5.2.7)). The minimum value of the threshold r_c is 9, which corresponds to the limit $b \to 0$.

In the unstable domain, Risken and Nummedal (1968a and b) found pulse-train solutions characterized by a dependence on $t - x/v$:

$$|\hat{E}| = \hat{E}(t - x/v); \quad |\hat{P}| = \hat{P}(t - x/v); \quad \hat{D} = \hat{D}(t - x/v) \qquad (9.1.12)$$

where v is the propagation velocity of the pulses. An integral number of equally spaced pulses can be present within the laser cavity. Figure 9.1.2 shows an example of this kind of behavior (clearly this behavior can be regarded as a special case of "passive" mode-locking). As in other cases of pulsing behavior, the population inversion D can take on negative values just after the passage of each pulse of the field. This regular pulsing occurs because the phases of all the excited cavity modes are stable and are phase-locked to the resonance mode. The solutions of the type (9.1.12) obey ordinary differential equations isomorphic with the Lorenz system (eq. (2.1.1)) (Ackerhalt et al., 1985), so that similar behavior is to be expected.

More recently, Lugiato and co-workers (Lugiato, Narducci et al., 1985; Lugiato and Narducci, 1985; Narducci et al., 1986) further analyzed multimode instabilities in homogeneously broadened ring lasers. On the one hand, they took into account the influence of a finite mirror reflectivity and allowed for an arbitary value of the gain parameter (i.e. they considered the fact that the field is not exactly uniform along the laser cavity in the steady state) (Lugiato, Narducci et al., 1985), and also considered the influence of an arbitary detuning between the atomic line and the reference cavity mode (Narducci et al., 1986). On the other hand, they demonstrated the existence of a close and useful link between single-mode and multimode instabilities in lasers and

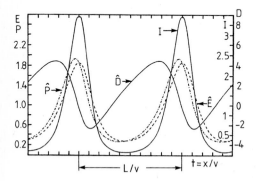

Fig. 9.1.2. Pulse-train solution predicted by eqns. (9.1.3) for $\gamma_\parallel = \gamma_\perp/2$, $\Lambda = 15$ and $L/v = 2\pi/(3.47\,\gamma_\perp)$. The direction of propagation is from right to left along the x-axis. The field intensity is represented by $I = |\hat{E}|^2$. (From Risken and Nummedal, 1968a and b.)

laser-related optical systems (Lugiato and Narducci, 1985). Their main conclusion was that the existence of a single-mode instability (for instance, in "bad-cavity" conditions) necessarily implies a corresponding multimode instability in the "good-cavity" limit and, conversely, the existence of a multimode instability in the "good-cavity" limit implies, under an appropriate additional condition, the existence of a single-mode instability for a sufficiently "bad" cavity. This link can be established not only for homogeneously broadened lasers but also for inhomogeneously broadened lasers, lasers with a non-uniform transverse intensity profile, lasers with saturable absorbers, laser systems in a Farbry-Pérot cavity and lasers whose dynamics are governed by rate equations.

Brunner et al. (1988) numerically found deterministic chaos in the time evolution of the complex electric field amplitude of a multimode gas laser (with a number of modes up to 13) in the presence of strong inhomogeneous line broadening. Regular behavior was found in the case of pure homogeneous broadening, indicating that the occurrence of irregular behavior requires that mode coupling due to large homogeneous broadening be counteracted, to some extent, by a mechanism that favors mode coexistence. The case of a multimode laser with an injected signal was analyzed theoretically by Narducci et al. (1985).

As we have already mentioned, a surprising aspect of the Risken-Nummedal-Graham-Haken multimode instability is that it has never been clearly identified in an experiment, in spite of its fundamental character. Probably other phenomena screen or prevent the appearance of this instability. This seems to be the case of the experiment performed with a dye laser that is described in the next section.

Table 9.1.1 gives a few laser systems with their parameters at which the multi-longitudinal mode homogeneously broadened laser instability is expected to occur. It is apparent that in some cases the parameters are by no means in an exotic range and the instability should therefore be observable in these lasers.

Table 9.1.1. Some typical requirements for a homogeneously broadened laser resonator for the appearance of multilongitudinal mode instability. It is assumed that a high finesse is achievable, which means low threshold.

Medium	$\gamma_\parallel$ [s^{-1}]	$\gamma_\perp$ [s^{-1}]	Resonator length [m] required	Typical resonator length [m]
CO_2	10^4	10^8	$400-600$ (!)	$1-2$
GaAs	10^9	10^{12}	$2 \cdot 10^{-2}$	10^{-4}
Dye	10^{10}	10^{12}	$4 \cdot 10^{-2}$	0.5
Na$_2$	10^8	10^8	4	0.5
YAG	10^4	10^{12}	5	0.5

9.1.2   Dye Lasers

An experiment has been conducted with a homogeneously broadened dye laser (Hillman et al., 1984) which was thought to provide the first observation of the Risken-Nummedal-Graham-Haken multimode instability described in the preceding section. In this experiment a continuous dye laser pumped by an Ar^+ laser with a resonator of low loss was used.

The principal observation is that the laser, which at low pump power emits in a few modes (or one) near the gain line maximum, splits its emission spectrum at higher pump power into two groups of closely spaced modes. The frequency separation of the two mode groups is very large and scales as the square root of the laser intensity, i.e. as the laser field and the laser transition Rabi frequency (Fig. 9.1.3). No central mode (group) is observed in the experiment.

However, attempts to describe this experiment theoretically within that context failed (Hillman et al., 1984; Lugiato et al., 1985; Narducci et al., 1986; Hillman and Koch, 1986). The correct explanation seems to have been found more recently by Fu and Haken (1987 and 1988). The distinctive feature of their model is that they take into account the presence of the large number of vibronic sublevels that exist in the singlet electronic ground state of a dye molecule. They do it in a simplified form by considering a "two-level" model in which the lower "level" consists of a large set of equally (and closely) spaced sublevels.

Since the contribution of each sublevel to the lasing process is determined by the amplitude of the corresponding polarization, Fu and Haken studied how the polari-

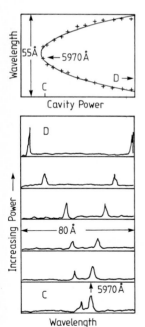

Fig. 9.1.3. Pump power dependence of a dye laser with high intensity. Spectra for different pump power and the wavelength dependence of the modes are shown. The parabolic dependence of wavelength on resonator power suggests a Rabi splitting as the mechanism (From Hillman et al., 1984.)

zations are distributed on the sublevels. Figure 9.1.4 shows an example of this distribution when the equations yield a two-frequency solution. It has been found that in the cases of Fig. 9.1.4(b)−(d), where the distribution shows two peaks, two frequencies are indeed emitted. Correspondingly, in (a) (only one peak) a single-frequency solution is favored, and in (e) a four-frequency solution would be favored.

The physical reason why a peak in the polarization distribution on the sublevels $|P_n|$ bifurcates when the pump power is sufficiently strong (this occurs, for instance, in Figs. 9.1.4.b and d) can be explained in the following way. Each component of the lasing field burns a hole in the distribution of the population inversions D_n (D_n represents the population difference between the upper level and the n sublevel of the lower level) by stimulated emission and hence modifies the distribution $|P_n|$. When this hole is burned so deeply that it cannot be compensated by the band shape, which is conducive to single-frequency operation, the peak will bifurcate. In accordance with the experiments (Fig. 9.1.3), the frequency separation between the two field components of the two-frequency solution looks like a parabola as a function of the pump power. However, the two-frequency solution has nothing to do with Rabi oscillations, because the population inversions D_n do not oscillate at the intermode frequency; rather, they are almost constant with time.

This model also explains several further features of the above-mentioned experimental results (Hilman et al., 1984). Further experiments on this dye laser system and refined models are currently being worked on.

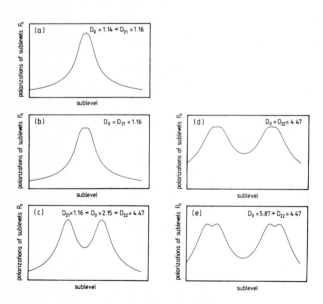

Fig. 9.1.4. Distribution of the polarizations $|P_n|$ on the ground sublevels in the two-frequency solution. D_0 represents the pump power. At $D_0 = D_{21}$ (b) the single peak splits into two peaks, and at $D_0 = D_{22}$ (d) each of the peaks splits into two new peaks. (From Fu and Haken, 1988.)

A completely different approach to chaos in multimode lasers was followed by Atmanspacher and Scheingraber (1986). It is known that, e.g., in multimode dye lasers, although the total output power is reasonably constant with time, the power in an individual mode fluctuates strongly. The main interest in this mechanism comes from the application of intracavity laser spectroscopy which uses multimode dye lasers. It is plausible that there should be mode-mode interactions since the dye medium is largely homogeneously broadened. Such interactions can, among others, lead to passive mode-locking, so it is not out of question that such interactions might also lead to chaotic dynamics.

Intensities of individual modes (or groups of modes) were recorded as a function of time and, with these intensity-time series, tests on fractal dimensions were made. Figure 9.1.5 shows the dependence of the number of points on the n-dimensional sphere radius in the artificial phase space constructed by the well known delay technique (see Chapter 4). The slope of the growth of the number of points inside a sphere of radius r with increasing r is given for different dimensions up to 20. "Plateaux" at slope > 2 appear. This can be interpreted by the phase-space of the system being three-dimensional, i.e. only three variables describe the dynamics of the laser, in spite of the fact that it emits on several thousand modes. As the variables, two fields and the inversion are suggested.

The interpretation is then that the several thousand modes are coupled together in two groups — the mechanism probably being spatial hole-burning — so that a situation very similar to mode-locking exists. The three variables then provide sufficient phase-space dimension to admit chaotic dynamics, which consequently produces the strong fluctuations of the individual mode intensities.

From this result, one may cautiously infer that the fluctuations of multimode lasers, which are a typical phenomenon, have at their roots simple but chaotic dynamics. Since the random fluctuations are then not caused by the great complexity of the system and various noise sources but rather by fairly simple dynamics, it would seem possible to control the mode intensity fluctuations by changing a laser parameter to, e.g., the steady-state conditions.

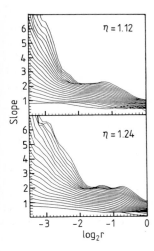

Fig. 9.1.5. Dimension plots for the emission of a multimode dye laser. Plateaux are found for different pump rates near dimension 2. (From Atmanspacher and Scheingraber, 1986.)

9.2   Counter-Propagating Modes

Let us consider again a homogeneously broadened two-level ring laser. A large literature on this type of laser exists because the ring geometry avoids "spatial hole-burning" and is therefore of advantage if single-mode laser emission is required. In this application it is usually attempted to suppress one of the two possible emission directions by a non-reciprocal-loss element within the resonator. Hence the instabilities of this system are those of the usual homogeneously broadened laser (the Lorenz model).

An important application, however, exists for a ring laser which emits on two counter-propagating modes. This is the "optical gyroscope". Rotation of the ring around the normal of the ring plane creates a phase asymmetry between the two emission directions owing to the (relativistic) Sagnac effect (Jacobs et al., 1984). This phase asymmetry obviously results in a difference of the emission frequencies of the two directions of laser emission. The latter can be easily measured by heterodyning of the two modes on a photon detector, thus permitting measurement of absolute rotation or angle. The "optical gyroscope" is used to replace the mechanical gyroscopes used so far.

Owing to the importance of this application and the fact that the two-mode ring laser, in particular its homogeneously broadened version, presents one of the experimentally and theoretically simplest cases in laser dynamics, this system has recently been studied with respect to its dynamical properties. The system chosen for the study is the CO_2 laser, homogeneously broadened with appreciable gain and well known molecular dynamics and spectroscopy, thus well suited as a model system.

As has been explained in previous chapters, the models generally have to consider polarization of the medium, inversion and fields. If certain of the corresponding relaxation rates are large in comparison with the others, the respective variables may be adiabatically eliminated. Thus for a dye laser, for example, the resulting equations describe only the two interacting mode fields. In this model it can be shown that the uni-directional emissions in both directions are stable (Mandel et al., 1984). Switching between them has been found to occur (Lett and Mandel, 1985) as a result of noise in the system or back-scattering or reflection from an external mirror. The solution in which the laser emits simultaneously in both directions is always unstable, which means that after a transient phase the laser settles into one of the uni-directional emission modes.

A CO_2 laser, on the other hand, allows only the medium polarization to be adiabatically eliminated and is therefore described by a larger number of equations (seven real equations, as is shown below). Consequently, a more complicated dynamical behavior is expected. Let us consider this case in some detail.

9.2.1 CO_2 (and Solid-State) Lasers

Initially, an experiment on a CO_2 ring laser emitting in two directions was carried out (Lippi et al., 1985). Figure 9.2.1 shows in a phase diagram in the laser discharge current

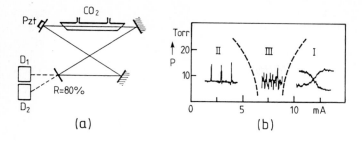

Fig. 9.2.1. (a) Experimental arrangement of bidirectional CO_2 ring laser. (b) State diagram of the CO_2 ring laser in (a). The discharge current (mA) and the helium partial pressure of the active gas are varied: (I) mode alternation; (II) self Q-switching; (III) irregular pulsing. (From Lippi et al., 1985.)

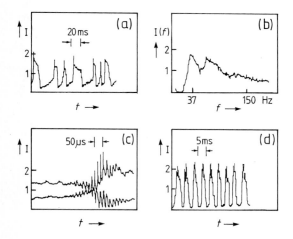

Fig. 9.2.2. Region I of Fig. 9.2.1. Irregular mode alternation (switching). (a) Laser intensity for one direction as a function of time. (b) Corresponding spectrum is continuous, indicating chaotic dynamics. (c) Mode-switch induced relaxation oscillations. (d) Numerical result of a model calculation to be compared with (a). From (Lippi et al., 1985.)

— laser gas pressure plane three different regions of qualitatively different dynamical laser behavior. In region I the laser emits in one direction stably when the laser resonator is tuned less than 0.2 homogeneous linewidths away from the line center. For larger detuning the laser spontaneously switches between the emission directions.

Figure 9.2.2 shows the laser output power as a function of time in one direction. Relatively sharp switches occur. As one might expect, the sharp step in the intensity causes "ringing" at the laser relaxation oscillation frequency. It must be noted, however, that these oscillations do not have the usual form of damped ringing following the switch. Rather, during the increase in the intensity the oscillation amplitude increases (Fig. 9.2.2c).

After reaching the maximum, the oscillation amplitude decreases. The opposite direction emits complementarily. During its switch-off phase, the relaxation oscillation amplitude increases (Fig. 9.2.2c) and decreases after the minimum intensity has been reached. The relaxation oscillations in the two directions are 180° out of phase. The growing and the damped spirals differ in frequency (theoretically by $\sqrt{2}$ (Mandel and Abraham, 1984)).

In region II, as opposed to the almost "square-wave" switching in region I, very sharp pulses develop spaced regularly in time (Fig. 9.2.3). These might develop out of the highest spike in the switching-induced relaxation oscillations. In any case they are reminiscent of the Q-switch pulses encountered in the case of a laser with an injected signal (see Chapter 7). One might argue that the bidirectional laser constitutes a similar situation: one mode locking the other by Bragg scattering from the spatial grating formed in the population inversion by the interference of the two modes. As in the case of the Q-switching of the laser with an injected signal, the Q-switch can be accompanied by damped relaxation oscillations (Fig. 9.2.3c) which are 180° out of phase for the two emission directions.

In region III sharp pulses occur in a chaotic fashion (Fig. 9.2.4). An increase in gain leads to increasingly irregular pulse heights and time between the pulses.

A detailed model has been set up recently to explain and to predict experimental results on CO_2 and solid-state lasers (Zeghlache et al., 1988; for previous studies, see the complete list of references cited therein). This model uses mainly two approximations:

1. The medium polarization is adiabatically eliminated.

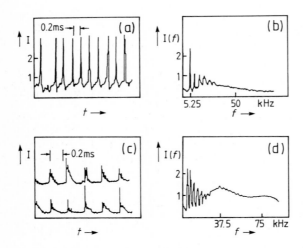

Fig. 9.2.3. Region II of Fig. 9.2.1 (experimental). (a) High-intensity pulses simultaneously in the two modes; (b) corresponding spectrum; (c) both modes shown simultaneously at different pressure/current. (d) Damped relaxation oscillations are discernible. (From Lippi et al., 1985.)

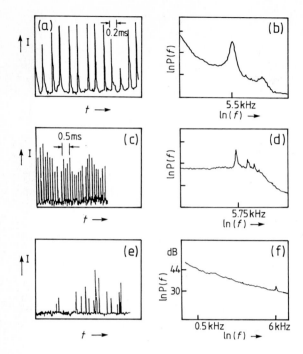

Fig. 9.2.4. Region III of Fig. 9.2.1 (experimental). Pulsing becomes more irregular with increasing current: (a) 5 mA; (c) 5.5 mA; (e) 6 mA; (b), (d), (f) corresponding intensity spectra of laser output. (From Lippi et al., 1985.)

2. The spatial modulation of the inversion created by the interference of the two counter-propagating waves is considered to be sinusoidal; higher order spatial Fourier terms are neglected.

Further, as in all models so far, the waves are considered to be plane and uniform in space. With these approximations, the set of ordinary differential equations describing the system consists of one equation for each field mode, one for the population inversion grating and one for the uniform population inversion. Since detuning is allowed for, fields and the inversion grating are complex, so that a set of seven coupled real equations have to be solved. Following Mandel and co-workers, these equations can be put into the form

$$
\begin{aligned}
\partial_\tau \hat{E}_1 &= -E_1 + (1 + i\varDelta)\tilde{A}(E_1 \mathscr{D}_0 + E_2 \mathscr{D}_1^*) \\
\partial_\tau E_2 &= -KE_2 + (1 + i\varDelta)\tilde{A}(E_2 \mathscr{D}_0 + E_1 \mathscr{D}_1) \\
\partial_\tau \mathscr{D}_0 &= -d_\parallel(\mathscr{D}_0 - 1) - \tilde{d}_\parallel \mathscr{D}_0(|E_1|^2 + |E_2|^2) - \tilde{d}_\parallel(E_1 E_2^* \mathscr{D}_1 + \text{c.c.}) \\
\partial_\tau \mathscr{D}_1 &= -\tilde{d}_\parallel \mathscr{D}_1 - \tilde{d}_\parallel \mathscr{D}_1(|E_1|^2 + |E_2|^2) - \tilde{d}_\parallel E_1^* E_2 \mathscr{D}_0
\end{aligned}
\tag{9.2.1}
$$

where $\tau = \kappa_1 \cdot t$, $\partial_\tau \equiv \partial/\partial\tau$, κ_i is the cavity damping rate for the field mode i ($i = 1$ and 2 for the forward and backward fields of complex amplitudes E_1 and E_2, respectively), $K = \kappa_2/\kappa_1$, $\Delta = (\omega_c - \omega_A)/\gamma_\perp$, ω_c is the nearest cavity resonance frequency (the same for both counter-propagating modes), ω_A is the atomic transition frequency, $\tilde{A} = A/(1 + \Delta^2)$, A is the usual pump parameter, $\mathscr{D}_0$ and $\mathscr{D}_1$ are the lowest order harmonic expansion terms of the population difference $D(x,\tau)$:

$$\mathscr{D}_n(\tau) = \frac{1}{L}\int_0^L D(x,\tau)e^{i2nkx}\,dx \tag{9.2.2}$$

(where n is an integer, x is the coordinate along the cavity axis and L is the cavity length), $d_\parallel = \gamma_\parallel/\kappa_1$ and $\tilde{d}_\parallel = d_\parallel/(1 + \Delta^2)$. Both field and inversion variables are dimensionless quantities normalized in a common way similarly to eq. (7.1.4) (Mandel and Agrawal, 1982; Zeghlache et al., 1988); also, as usual, a factor $\exp(-i\omega_c t)$ has been omitted in E_1 and E_2. For purpose of comparison, we can say on the one hand that the set of eqns. (9.2.1) is the generalization of eqns. (7.1.3) or (7.1.5), which are valid for a homogeneously broadened, single-travelling wave, resonantly tuned class B laser, to the case of two counter-propagating modes and any cavity tuning Δ. On the other hand, they do not display an x-dependence and a spatial derivative $\partial/\partial x$ (as they appear, for instance, in eqns. (9.1.3) for the case of co-propagating modes) because they have been eliminated through the harmonic expansion (9.2.2) (at the cost of increasing the number of variables) and through the decomposition of the total field into its counter-propagating components E_1 and E_2.

The analysis by Zeghlache and co-workers (1988) concentrated on the stability of the steady-state solutions and on the intensity pulsations of the time-dependent solutions. This study was later completed (Hoffer et al., 1988) with studies of the behavior of the frequencies and phases of the two modes. To this end the fields and inversion grating were expressed in terms of real amplitudes and phases:

$$E_1 = X_1 e^{i\phi_1}, \quad E_2 = X_2 e^{i\phi_2}, \quad \mathscr{D}_1 = Y e^{i\psi}. \tag{9.2.3}$$

The results obtained in these studies (Zeghlache et al., 1988; Hoffer et al., 1988) qualitatively confirm the principal features of the above-described experimental observations on a CO_2 laser. These results can be summarized as follows:

(i) The main steady states are the uni-directional forward and backward solutions:

$$I_1 = |E_1|^2 = A - (1 + \Delta^2), \quad E_2 = 0 \tag{9.2.4}$$

which exists for $A \geqslant 1 + \Delta^2$, and

$$I_2 = |E_2|^2 = \frac{A}{K} - (1 + \Delta^2), \quad E_1 = 0 \tag{9.2.5}$$

which exists for $A \geqslant K(1 + \Delta^2)$. If we choose $\kappa_2 \geqslant \kappa_1$ (i.e. $K \geqslant 1$), we may denote eq. (9.2.4) as the "strong-mode" solution and eq. (9.2.5) as the "weak-mode" solution. For central resonator tuning both solutions are stable and switching between the emission directions may only occur due to noise in the system.

(ii) The instabilities reported in these investigations only appear when a very small but non-zero detuning exists (for instance, $\Delta \gtrsim 0.05$), for almost any value of A above the laser threshold. They arise from Hopf bifurcations of the steady states when either Δ or A is increased.

(iii) The variable $\mathscr{D}_1$ describes a longitudinal spatial grating in the population inversion, which appears as a result of the interference of the counter-propagating fields and plays a key role in the dynamics of the system. One field can scatter from the grating into the other field, giving rise to additional coupling between them, beyond simple cross-saturation of the gain. The grating is time dependent, so that its dynamics can affect the stability of the laser operation. On the one hand, the slow relaxation rate of the population inversion allows the grating to persist long after the decay of the fields that created it, and on the other, the grating has a complex (i.e. phase-shifted) amplitude, which can make the scattered fields interfere constructively or destructively with the incident fields.

(iv) The stability of a given steady state is determined by two factors which are related to the stability of the emitting mode of the field (coupled to the average population inversion) and the stability of the suppressed mode which has zero intensity (coupled to fluctuations in the zero-amplitude population grating). The strong mode has the stability of the usual single-mode laser in the rate equation limit, that is, it is always stable, although with damped relaxation oscillations for suitable parameters. This means that destabilization occurs only through a suitable perturbation of the field of the suppressed mode and the grating of the population inversion in the presence of a sufficient detuning. As all the seven variables are coupled by non-linear terms, the growth of the grating and the weak-mode field ultimately destabilize the strong mode as well.

(v) Figures 9.2.5 and 9.2.6 show the evolution with time of the intensity of each mode (a) and of the total phase $\mu = \phi_1 - \phi_2 + \Psi$ (b), which determines whether the contribution of the field scattered by the grating leads to growth or decay in the amplitudes (Hoffer et al., 1988). The values of the parameters are typical for a CO_2 laser (Lippi et al., 1985). For small cavity detuning ($\Delta = 0.1$) the system switches regularly from forward to backward emission (Fig. 9.2.5.a). The time between consecutive switches increases with decreasing detuning (critical slowing down) and it shows $10-20\%$ fluctuations whose origin probably lies in deterministic chaos. A shift of $\pm \pi$ rad as either mode passes through its minimum is observed on the combined phase μ (Fig. 9.2.5.b). The frequency of each mode remains constant (it coincides with the frequency of the steady-state solution, Δ) when it dominates, but it is pulled or pushed by about 25% during growth and decline.

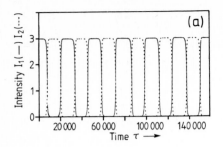 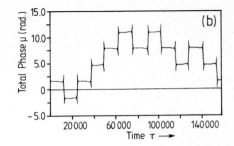

Fig. 9.2.5. (a) Evolution with time of the intensities I_1 (strong forward mode) and I_2 (weak backward mode). (b) Evolution with time of the total phase $\mu = \Phi_1 - \Phi_2 + \Psi$. Parameter values appropriate for a CO_2 ring laser: $d_\| = \gamma_\|/\kappa_1 = 1.7857 \times 10^{-4}$, $K = \kappa_2/\kappa_1 = 1.00007$ (i.e. the strong and weak modes have almost the same intensity), $A = 4.0$ and $\Delta = 0.1$ (i.e. small cavity detuning). (From Hoffer et al., 1988.)

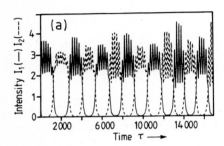 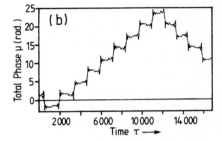

Fig. 9.2.6. Results as in Fig. 9.2.5, for $\Delta = 0.4$. (From Hoffer et al., 1988.)

For larger cavity detuning ($\Delta = 0.4$), the relaxation oscillation after the switching process, which was very small (hardly visible) in Fig. 9.2.5, appears much more pronounced (Fig. 9.2.6.a). For small detunings the relaxation oscillation frequencies of the mode switching on and of the mode switching off differ by a factor of $\sqrt{2}$ as in the resonant case (Mandel and Abraham, 1984). The combined phase μ still shows jumps by $\pm \pi$ rad as either mode passes through its minimum intensity.

For still larger cavity detunings, the depth of modulation of the mode that emits increases. For $\Delta \geqslant 0.6$ the square-wave mode of switching breaks down entirely in favor of pulsations, which are generally irregular and chaotic except for some windows of periodic oscillations. Strong transfer of energy between the counter-propagating modes continues to be apparent in any case. For more details, see Zeghlache et al. (1988) and Hoffer et al. (1988).

For the case of solid-state lasers, which are also class B lasers, specific features have been investigated by Khanin and co-workers (Khanin (1988) and references cited therein). In the case of resonator asymmetry, the two weak-mode relaxation oscillation frequencies may differ. It was proved experimentally on a solid-state laser with a phase-asymmetric resonator that all three relaxation resonances occur (Fig. 9.2.7).

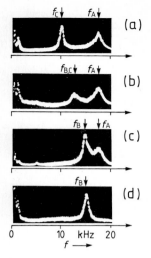

Fig. 9.2.7. Intensity fluctuation spectra of a bidirectional Nd-YAG laser. (a–c) Under different conditions three resonances, F_A, F_B, F_C (driven by noise), are observable. (d) Weak mode emission. (From Khandokhin and Khanin, 1985.)

9.2.2 Optically Pumped Bidirectional Ammonia Laser

Experiments on bidirectional ring lasers have also been performed on optically pumped ammonia lasers in the far-infrared region (see Chapter 8) (Klische and Weiss, 1985).

A detailed account of the similarities of the dynamics observed on ammonia lasers and the CO_2-laser model was given by Abraham and Weiss (1988). If the pump is tuned to the ammonia absorption line center, uni-directional emission of the ammonia ring laser is not possible. This system differs from the laser with adiabatic elimination of polarization discussed above, since the relaxation times of field, polarization and inversion are all comparable. Nonetheless, qualitatively similar dynamics are found, as observed so far.

The two kinds of laser systems appear qualitatively similar in that mode changing and relaxation oscillations are basic dynamic mechanisms. These have different characteristic frequencies.

A typical dynamics is mode switching with relaxation oscillations triggered by the switches. If the damping time of the relaxation oscillations is larger than the switching time, the relaxation frequency becomes effectively undamped.

Intensity spectra of this mode switching are shown in Fig. 9.2.8: (g) shows the switching frequency (f_2) about 1/4 of the relaxation oscillation frequency, (h) shows f_1 and f_2 phase-locked at an exact ratio $f_1/f_2 = 4$; (e) and (f) show the same case for a ratio near 3 (e) and locked exactly at 3 (f), and (a) shows a locked state in which $f_1/f_2 = 2$; the latter correspondingly has the appearance of a period-doubled emission spectrum.

The various ratios of f_1/f_2 were chosen by tuning the ammonia laser resonator. Locking occurred when the pump power was increased (or the frequency f_1 was chosen to be near a multiple of f_2). A further increase in pump power after the locking would,

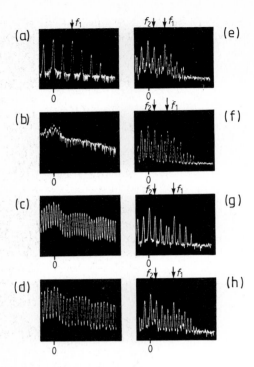

Fig. 9.2.8. Self-pulsing of a bidirectional ammonia ring laser. Measured intensity spectra. (a) Pulsing with period doubling. (e) Pulsing at two frequencies, f_1, f_2, which can lock at a ratio of (f) 3 or (g) and (h) 4. Subsequent to the locking, period doubling, visible in (f) and (h), leads to chaotic spectra like (b). The well known "period-three" and "-five" window in the chaotic range ((d) and (c)) are also observed. (From Klische and Weiss, 1985.)

irrespective of the f_1/f_2 ratio, start a period-doubling cascade to chaos. The first sub-harmonics are clearly visible in Fig. 9.2.8 (f) and (h), the locked states. The chaotic spectrum following the period-doubling cascade is shown in Fig. 9.2.8 (b). The well known large periodic windows of period three and five in the period-doubling chaos were also clearly observed (Fig. 9.2.8 (c) and (d)) (Klische and Weiss, 1985).

9.2.3 He-Ne Ring Laser

On an inhomogeneously broadened (He-Ne) ring laser which also has relaxation rates of field, polarization and inversion of comparable magnitude, bidirectional emission also results in dynamics (Danileiko et al., 1986) when the two counterpropagating modes are coupled by an external mirror.

Figure 9.2.9 shows spectra taken at fixed conditions (gain, coupling) when the laser resonator length is varied. Period-one, -two and -four and chaotic spectra were ob-

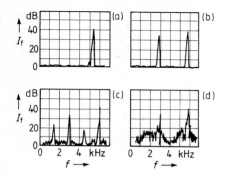

Fig. 9.2.9. Intensity spectra of bidirectional ring He-Ne laser when the laser resonator frequency is varied. (a) Periodic pulsing; (b) period doubling; (c) period quadrupling; (d) chaos. Detunings: see text. (From Danileiko et al., 1986.)

served for laser frequency detuning from the line center by 0, 9, 13 and 17 MHz, respectively. Another pulsing regime showing period-two and -three subharmonics and chaos was found 100 MHz away from the gain line center, whereas 150 MHz away at the highest gain a subharmonic cascade of period one, two, three, four, five, ..., chaos was found. It was stated that the period-doubling cascade near the line center was also found in a system of equations describing the laser that involve only the two fields and their phase difference.

9.3 Other Cases

9.3.1 Multimode Laser with Intracavity Frequency Doubling

A system related to the homogeneously broadened two-counter-propagating mode laser is a laser with an internal frequency-doubling crystal operating on two (or more longitudinal) modes. This scheme is used in YAG lasers which are diode laser pumped, and the output of which is to be frequency doubled in order to achieve visible (green) light of substantial power with a long-lived laser device. The cavity mirrors are reflecting for the laser radiation and transparent for the frequency-doubled radiation.

The experiments with a multimode YAG laser pumped longitudinally by a 200 mW, 808 nm laser diode with a KTP crystal in a high-finesse laser resonator (Baer, 1986) showed an erratic output power (Fig. 9.3.1) with almost 100% intensity fluctuations on a 10 μs time scale. Chaotic dynamics was found to be at the root of these fluctuations. If more than one mode is oscillating, the KTP crystal not only produces the harmonic frequency of each mode but also the sum frequency of the two modes, which strongly couples the modes.

The equations describing this system, e.g. in a case with N longitudinal modes, are (Baer, 1986):

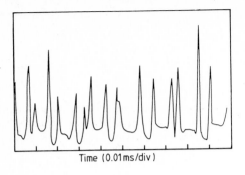

Time (0.01ms/div)

Fig. 9.3.1. Chaotic output pulsing of an Nd-YAG laser with an internal frequency doubler. (From Baer, 1986.)

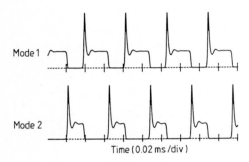

Mode 1

Mode 2

Time (0.02 ms/div)

Fig. 9.3.2. Calculated oscillations of the mode amplitudes for the two-mode laser. (From Baer, 1986.)

$$\frac{\mathrm{d}I_i}{\mathrm{d}t} = \frac{1}{\tau_c}\left(G_i - \alpha_i - \varepsilon I_i - 2\sum_{j\neq i}^{N}\varepsilon I_j\right)I_i$$

$$\frac{dG_i}{dt} = -\frac{1}{\tau_f}\left(\beta I_i + \sum_{j\neq i}^{N}\beta_{ij}I_j + 1\right)G_i + G_i^{\mathrm{o}}$$

(9.3.1)

where $i,j = 1,2,\ldots, N$ denote each mode, the variables I_i and G_i represent the field intensity and gain for the mode i, respectively, τ_c is the cavity round-trip time (0.5 ns), τ_f is the fluorescence lifetime (0.24 ms), α_i is the mode-i losses (0.015), ε is the non-linear coupling coefficient due to the generation of the green light (5×10^{-5} W^{-1}), β is the saturation parameter (1 W^{-1}), $\beta_{ij} = \beta_{ji}$ is the cross-saturation parameter describing the influence of mode j on the gain of mode i (0.666 W^{-1}) and G_i^{o} is the small signal gain for mode i (0.12). The numbers in parentheses are typical values for the laser diode-array-pumped Nd:YAG laser described (Baer, 1986). These equations are written in the rate-equation approximation and describe the frequency doubling of each mode and the sum-frequency generation processes through effective intensity-dependent losses (see terms $-\varepsilon I_i$ and $2\sum_{j\neq i}^{N}\varepsilon I_j$, respectively, in the first equation of (9.3.1).

A numerical integration in the case of two modes (i.e. $N = 2$) shows a periodic alternating switching between modes 1 and 2 (Fig. 9.3.2). The excellent agreement with observations is apparent from Fig. 9.3.3. The dynamics can easily be visualized:

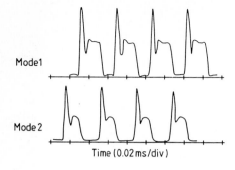

Mode1

Mode2

Time (0.02 ms/div)

Fig. 9.3.3. Experimentally observed pulsing for the two-mode laser. Compare with Fig. 9.3.2. (From Baer, 1986.)

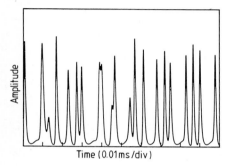

Amplitude

Time (0.01 ms/div)

Fig. 9.3.4. Calculated pulsing of three-mode laser, to be compared with Fig. 9.3.1. (From Baer, 1986.)

1. Mode 1 on steady-state emission.
2. Mode 2 emits a pulse overshooting the steady-state value, thereby increasing the loss of mode 1 (sum-frequency generation) so that mode 1 goes below the threshold.
3. Mode 2 stabilizes after some relaxation oscillations to its steady-state value, stabilizing the loss for mode 1. The gain of mode 1 (pumping) increases until mode 1 reaches the threshold.
4. Mode 1 emits a pulse, overshooting so that mode 2 goes below the threshold as in 1. But mode 1 and 2 exchanged. And so on.

Figure 9.3.4 shows the intensity fluctuations for three modes and this compares well with Fig. 9.3.1 for measurements on the laser.

From the above it is clear that a single-mode laser with internal frequency doubling should be stable, since its phase-space dimension is too small to admit chaos (for $N = 1$ one has only two equations). This is what is observed experimentally. It follows that satisfactory generation of green light from diode laser-pumped AG lasers will require measures that ensure single-mode operation.

9.3.2 Transverse Multimode Lasers

A number of experiments on multimode laser instabilities have been carried out. The conceptually clearest experiment made a connection with the decades-old experience that pulsed solid-state lasers tend to emit in irregular bursts of pulses (Hauck et al., 1983).

It was found that the pulsing behavior of a flashlamp-pumped neodymium laser depends sensitively on the Fresnel number of the resonator, adjusted by a variable iris inside the laser resonator.

A calculation taking into account the radial dependence of the gain saturation, transforming the field distribution on one mirror to that on the other mirror, using the full Kirchhoff diffraction integral repetedly, shows that in the early stages of the pump pulse the TEM_{00} mode starts. Increased pumping at later stages then creates sufficient inversion in the outer regions of the laser rod for the losses of higher transverse modes to be overcome.

The calculation shows first single-mode emission, then two-mode emission and finally three- or more mode emission, irregular in pulsing, which is interpreted as chaotic dynamics. Figure 9.3.5 shows a comparison of calculated pulsing, experimentally measured pulsing and a streak camera film of the lateral width of the mode pattern.

The temporal structure of the pulses calculated appears to reflect the observations fairly well. The mode pattern is seen to vary irregularly, as would be expected for a chaotic system. It should be mentioned that this work on the instability of the neodymium laser is the first one that has attempted to include the transverse mode structure. All other theoretical models use the uniform, plane-wave approximation.

A very similar observation was made on a continuously operating CO_2 laser which could show some relationship with the theoretical work by Shih and Milonni (1984), in which it was shown that spatially inhomogeneous pumping can lead to chaotic emission through a period-doubling route (Biswas and Harrison, 1985). The free spectral range of the usual CO_2 lasers is comparable to the gain linewidth. If the TEM_{00} mode is detuned from the gain line center, higher order transverse modes which have a frequency different from the TEM_{00} modes may lie closer to the line center and may therefore also oscillate. The successive onset of emission of several modes was observed

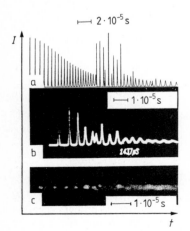

Fig. 9.3.5. Pulsing of a multimode (single longitudinal, multi-transverse mode) Nd-YAG laser with flashlamp pump. (a) Calculation of model; (b) experimental pulsing of laser intensity; (c) mode structure of laser. (From Hauck et al., 1983.)

by the beat frequencies between them (Biswas and Harrison, 1985). On resonator tuning, starting with the TEM_{00} mode on the line center, a second mode appears, producing the beat frequency shown in Fig. 9.3.6a. The beating between three modes observed further away from the gain line center is shown in Fig. 9.3.6b. In Fig. 9.3.6c and d, the irregularity is attributed to chaotic interactions. It is difficult to judge if these pictures show a real interaction and if the dynamics therefore are in fact chaotic, or if relatively complex beats of independent, non-interacting modes are observed. A spectrum is, of course, much more informative since it shows the harmonics and sum and difference frequencies characteristic of the non-linear interaction.

Similar observations were made on a high-gain He-Ne laser at a wavelength of 3.39 μm (Weiss et al., 1983). The laser frequency was controlled by a reference laser

500 ns
I I

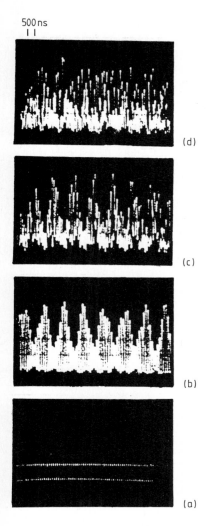

(d)

(c)

(b)

Fig. 9.3.6. Route to chaos of a multimode CO_2 laser (experiment). (a) Beating between two laser modes; (b) quasi-periodicity as a result of the interaction of three transverse modes; (c) onset of chaos; (d) fully developed chaotic pulsing. Tuning of laser resonator is varied. (From Biswas et al., 1985.)

(a)

stabilized to a methane Lamb dip (Fig. 9.3.7). As a control parameter simply the tilting angle of one of the resonator mirrors was used. A tilt increases the relative loss of the various transverse modes.

This is a complex system: in addition to the complication of standing waves and multitransverse modes, the laser medium is inhomogeneously broadened. In the experiment only the intensity modulation was detected and analyzed by a radio-frequency spectrum analyzer.

Figure 9.3.8 shows an intensity spectrum corresponding to three-mode emission, showing two beats around 60 MHz — the free spectral range of the laser resonator corresponding to three oscillating modes. The difference in the beat frequencies appears as a spectral line near zero frequency, generated by the non-linear interaction of the three modes ("beat-beat"). The "beat-beat" spectrum served for the observations.

Around the gain line center the three classical routes to chaos were found when the mirror tilting angle was varied by $2.5 \cdot 10^{-4}$ rad.

Figure 9.3.9a shows the well known period-doubling transition to chaos clearly identified by the successive appearance of the spectral lines at $m \cdot f/2n$ (where m and n are integers) and ending in the chaotic noise spectrum. Figure 9.3.9b shows the well known quasi-periodicity transition (Ruelle-Takens transition to chaos). The dynamics

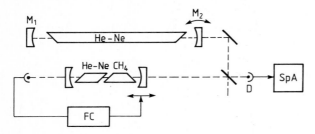

Fig. 9.3.7. Measurement set-up for observation of multimode chaos of a 3.39 µm He-Ne laser, consisting of a 2.5 m long He-Ne laser, heterodyned for precise resonator length control with a methane Lamb-dip-controlled frequency-stable second He-Ne laser. Mirror M_2 was tilted to observe different output pulsing. FC, methane-stabilized laser frequency control; D, photodiode; SpA radio frequency spectrum analyzer. (From Weiss et al., 1983.)

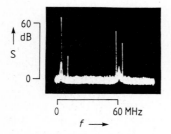

Fig. 9.3.8. RF spectrum of 3.39 µm He-Ne laser intensity showing two mode beats around 60 MHz (three-mode oscillation) and the "beat-beat" at low frequency. Its frequency equals the difference between the beats around 60 MHz. (From Weiss et al., 1983.)

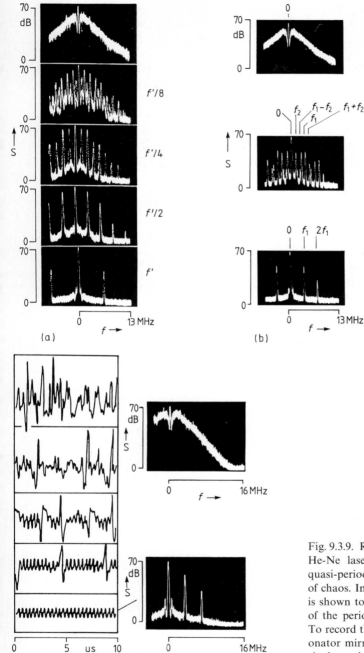

Fig. 9.3.9. Routes to chaos of multimode He-Ne laser. (a) Period doubling; (b) quasi-periodicity; (c) intermittent onset of chaos. Intensity as a function of time is shown together with intensity spectra of the periodic and fully chaotic cases. To record these sequences, one laser resonator mirror was tilted away from the single-mode emission alignment. (From Weiss et al., 1983.)

starts with one oscillation (f_1) (brought about by three-mode interaction) and a further mirror tilt adds another frequency (f_2) (possibly another mode) which interacts non-linearly with f_1 producing all linear combination frequencies of f_1 and f_2. A further mirror tilt then leads to chaotic emission.

Figure 9.3.9c shows the classical intermittent onset of chaos. A regular, although already non-linearly distorted, periodic oscillation starts the process. As can be seen from the spectrum, this is the same oscillation as that starting Fig. 9.3.9b. Tilting results in sudden bursts of large-amplitude pulses in an irregular fashion. With increasing tilt angle these bursts become increasingly frequent until the whole time evolution consists of irregular bursts. As the spectrum shows, this state represents fully chaotic dynamics.

Transverse effects and spatial instabilities are treated in detail in chapter 10.

9.3.3 Diode Laser with External Reflectors or Resonators

A diode laser differs from most other lasers in three respects: (1) it has a very strong spontaneous emission; (2) its material refractive index changes with light intensity; and (3) the mirror reflectivities are small, so the laser is relatively "open" to all perturbations from outside. As we have already mentioned in previous chapters, the relaxation rate of the polarization is fairly large compared with the field and inversion relaxation rates and the laser can therefore be treated with adiabatic elimination of polarization (class B laser).

Consequently, the diode lasers can exhibit damped relaxation oscillations which are driven by the strong spontaneous emission and lead to a considerable noise component at the relaxation oscillation frequency. Evidently externally coupled reflectors or resonators will cause additional noise frequencies or even undamped oscillations. Owing to the practial importance of noise enhancement by unavoidable external reflections, and also the amplitude and frequency noise reduction, possible when external resonators are tuned to the laser frequency, the system of a diode laser with some feedback from an external reflector has been studied in some detail.

The spontaneous emission from a diode laser is usually concentrated into the laser modes even by the relatively low-finesse laser resonator modes. It is evident that additional resonances may channel the spontaneous emission energy to other frequencies.

Thus typical resonances of the system are, for example:

(1) the laser mode frequency (single-mode solitary laser emission);
(2) the external optical resonator optical resonance frequencies;
(3) the difference of external resonator mode frequencies;
(4) the "compound" resonator (consisting of laser diode resonator plus external resonator) frequencies; and
(5) the relaxation oscillation frequency.

It is intuitively plausible that, when two of these resonance frequencies match, one might generally expect some "resonant destabilization", as in the case of the laser with two counter-propagating modes, each of which itself shows only damped relaxation oscillations.

Although the solitary laser is stable (as a class B laser), unwanted reflections into the laser or deliberate coupling with the aim of reducing frequency fluctuations to an external reflector (resonator) may lead to periodic or chaotic laser pulsations.

Experiments of various kinds have been conducted. The first (Glas et al., 1983) simply used a diode laser whose radiation was fed back into the laser by a reflection grating. The grating provides some spectral discrimination of diode laser modes other than the desired one. The threshold pump current is decreased when the laser is coupled to the grating owing to the additional reflectivity of the grating. For current values between the solitary laser threshold and the threshold for the laser coupled to the grating, irregular pulsing is observed.

The pulse trains observed were Fourier analyzed and indications of harmonics and two subharmonics were found, lending some credibility to the assumption that the observed irregular pulsing represents chaotic dynamics (Glas et al., 1983).

A calculation of the instability of a semiconductor laser has been given by Otsuka and Kawaguchi (1984). It was shown in a model that, whenever the difference in frequency between a mode of the diode laser and a mode of the external resonator comes close to the relaxation oscillation frequency of the laser or its subharmonic, the laser will emit sustained oscillations at the subharmonic or the relaxation oscillation frequency itself (see Figs. 9.3.10a−d).

If the frequency difference of a laser resonator and an external resonator is set slightly different from the relaxation oscillation frequency, an intermittently chaotic pulsing is found (Fig. 9.3.10e). One would expect to see such an intermittency close to all subharmonic-locking points; however, it may be imagined that it is more difficult to observe these near-locked chaotic states at the subharmonics since the locking ranges are smaller here.

An experiment similar to the model described has been reported by Mukai and Otsuka (1985). Figure 9.3.11 shows the experimental arrangement. The laser diode output is collimated with two lenses, one beam being reflected back from a mirror through an absorber to attenuate the reflected light. The output from the other side of the laser is recorded by a fast avalanche photodiode and by a spectrograph, the former to record the pulses of the intensity and the latter to observe the number of modes of the laser diode. From the intensity pulsing the intensity spectra were calculated for different currents.

Subharmonic pulsing of the external resonator mode difference frequency $C/2L$ is clearly observed. Figure 9.3.12 shows the sequence of spectra taken at increasing current showing that the subharmonic pulsing ends in a chaotic emission.

The model suggested to explain these pulsing instabilities involves the frequencies of the compound resonator (i.e. the laser diode coupled to the external resonator) and the modes of the external resonator alone.

It may be shown that the strong amplified spontaneous emission, if enhanced by resonances produced, e.g., by an external resonator, may even exceed the laser threshold (ASE modes). The resonance destabilization scheme is then that the difference in frequency between the ASE mode and the compound resonator mode can be close to a

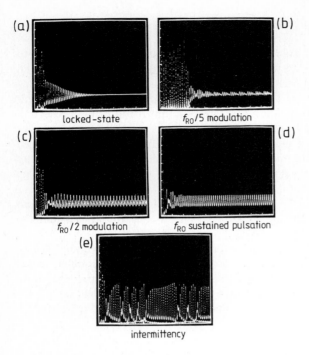

(a)   (b)

locked-state $f_{RO}/5$ modulation

(c)   (d)

$f_{RO}/2$ modulation f_{RO} sustained pulsation

(e)

intermittency

Fig. 9.3.10. Pulsing as calculated for a diode laser with an external reflector. Sustained oscillations at a subharmonic of the relaxation oscillation frequency arise whenever the difference between the laser resonator frequency and the external resonator frequency equals a subharmonic of the relaxation oscillation frequency. (a) No frequency difference between external and laser resonator; (b) frequency difference is $(1/5) \cdot f_{\text{Rel.Osc.}}$; (c) $(1/2) \cdot f_{\text{Rel.Osc.}}$; (d) $f_{\text{Rel.Osc.}}$; (e) intermittent behavior if frequency difference is not an exact subharmonic of $f_{\text{Rel.Osc.}}$. (From Otsuka and Kawaguchi, 1984.)

subharmonic of the external resonator mode frequency difference. The laser will then pulse at the corresponding subharmonic of the external resonator frequency difference.

If, for example, the detuning of the compound resonator mode from the ASE mode is half of the external resonator mode frequency difference, the pulsing will be at $1/2$ v_{ext}. On increasing the detuning to two thirds from the ASE mode ($1/3$ v_{ext} from the next ASE mode), the pulsing will be at $1/3$ v_{ext}, and so on at $1/4$ v_{ext}, $1/5$ v_{ext}, etc. When the detuning is almost 1 v_{ext} (almost at the next ASE mode) the locking is not complete, and an intermittent chaos is again to be expected. Thus a subharmonic cascade at v_{ext}/n is observed, ending in chaotic emission (and finally in v_{ext} fundamental pulsing). Again one may expect that the intermittency occurs at all subharmonics, even though it is probably more difficult to observe.

Spectra calculated from the model describing the interaction of ASE and compound resonator modes are shown in Fig. 9.3.13, giving evidence that this model agrees with the observations.

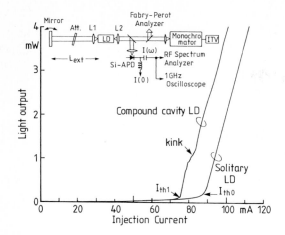

Fig. 9.3.11. Experimental set-up to record diode laser instabilities (see text) together with laser light output as a function of driving current (with and without external resonator). The dynamics occurs near the "kink". (From Mukai and Otsuka, 1985.) I_{th1}: threshold current without optical feedback; I_{th0}: threshold current with optical feedback.

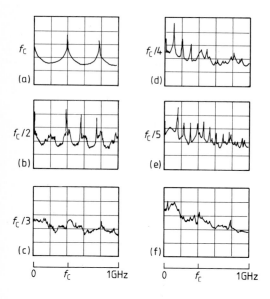

Fig. 9.3.12. Observed power spectra of Laser Diode with external reflector. (a) to (e) shows pulsing at subharmonics 1 to 5 resp. of the external resonator frequency. (f) shows chaotic emission (From Mukai and Otsuka 1985).

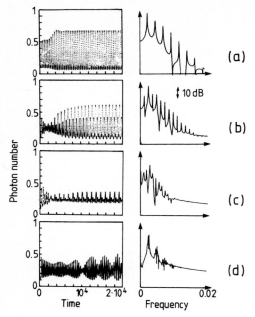

Fig. 9.3.13. Calculated laser pulsing from model. Periodic (a), period-doubled (b), period-tripled (c), and chaotic emission (d) are shown in the time picture and as spectra. Compare with Fig. 9.3.12. (From Mukai and Otsuka, 1985.)

In a further experiment (Cho and Umeda, 1986), the time-dependent output of a semiconductor laser with an external reflector was analysed in terms of fractal dimension. Figure 9.3.14 (a) shows the plot of the number of points within a sphere of radius d as function of sphere dimension. Figure 9.3.14 (b) shows a convergence of the dimension at 5.5, which is taken as further proof that the irregularly pulsing emission is originating from a strange attractor.

Semiconductor lasers can, like other lasers, be operated emitting repetitively pulsed radiation if a saturable absorber is inside the laser resonator. In GaAs semiconductor lasers this is achieved by damaging the active channel, e.g. by proton bombardment. Thus absorbers are created that can be saturated by the high light intensity in the laser waveguide ($10^6 - 10^7$ W/cm^2). The laser then emits short pulses (a few tens of picoseconds width) at a repetition frequency corresponding to the laser relaxation oscillation frequency. Kuznetsov and co-workers (1987) operated such a "passively Q-switched" semiconductor laser in an external resonator by antireflection coating one laser facet and reflecting the emitted light back into the laser channel from a $10 - 25$ cm distant mirror. The highly divergent output beam was, as usual, collimated by a lens of large numerical aperture.

It is found that the pulsing frequency locks to a multiple of the external resonator round-trip frequency. The multiple number gives the number of pulses propagating simultaneously inside the external resonator. This is a typical case of passive mode locking. Figure 9.3.15 shows the pulsation frequencies as a function of diode current. The fact that the pulsing frequency is locked to the resonator length of the external resonator results in hysteresis, which is shown experimentally in detail for the $m = 2$ and $m = 3$ states (Fig. 9.3.16).

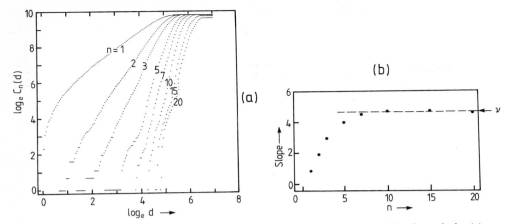

Fig. 9.3.14. Dimension test on a diode laser with an external resonator, pulsing irregularly. (a) Growth curves; (b) dimensions for different embedding space. The result suggests a dimension of ca. 5, for which no theoretical explanation is given. (From Cho and Umeda, 1986.)

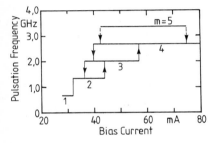

Fig. 9.3.15. Pulsing frequency of antireflection-coated self-pulsing laser diode in an external resonator. This can be described as passive mode locking. The numbers on the different branches are the pulses simultaneously traveling within the resonator. There is multistability (coexisting states). The state with five pulses in the resonator can only be reached by "hard" excitation. (From Kuznetsov et al., 1987.)

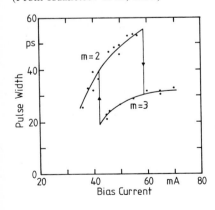

Fig. 9.3.16. Pulse width of the antireflection-coated self-pulsing laser diode in external resonator as a function of laser current, showing coexisting states. (From Kuznetsov et al., 1987.)

It is found that the $m = 5$ state, the state in which five pulses are simultaneously in the resonator, can only be reached by "hard" excitation, i.e. by a suitable choice of initial conditions. It can be reached experimentally, e.g., by blocking the resonator when the drive current is held above the $m = 3$ to $m = 4$ transition. This multistability of dynamic states (pulsing states with different numbers of pulses in the resonator) may allow interesting applications. Transitions between the states can be induced by light injection. A simple model (Kuznetsov et al., 1987) describes the phenomena qualitatively.

Similarly, pulsing of simple laser diodes without a saturable absorber has been observed with an antireflection-coated laser diode in a ring resonator (O'Gorman et al., 1988). This arrangement again produces pulsing at a frequency that is a multiple of the resonator round-trip frequency. The laser in this case is not self-pulsing. The pulsing probably results from mutual destabilization of the two counter-propagating modes as in Section 9.2. At the edges of the locking ranges, irregular pulsing is again observed, which is probably the intermittent type of pulsing always observed close to locking range edges.

9.3.4 Actively Mode-Locked Lasers

At is well known, active mode locking is widely used for short and ultrashort light-pulse generation (New, 1983; French et al., 1986). However, the instabilities arising in this regime have been investigated in less detail.

An analysis of mode-lock destruction in a cw solid-state laser with loss modulation has been reported by Golyaev and Lantratov (1983). A basic role in this process has been attributed to unlocking border modes that are due to detuning of the modulation frequency and beat frequency. Another investigation of the active mode-locking destruction process in a solid-state laser was reported (Bezaeva et al., 1985), in which a transition to chaos accompanied by low-frequency period doubling was observed.

Quasi-periodic oscillations and chaos have been observed in a gas-discharge He-Ne laser operating at 1150 nm (Melnikov et al., 1988). The destruction of active mode locking is shown here to be connected with the appearance of supermodes.

The pulse train envelope of a synchronously pumped mode-locked dye laser has been experimentally studied as a function of cavity length detuning (MacFarlane et al., 1988). Regular and erratic amplitude modulations were observed under different conditions.

10 Transverse Effects and Spatial Instabilities

In all preceding chapters, the calculations and the interpretations of observations have been made on the assumption that the laser field is confined between plane infinitely extended mirrors and that the medium extends laterally to infinity and is homogeneous throughout. Of course, no real optical device can realistically be described by such assumptions. E.g. for a transverse Gaussian field distribution, an averaging over the various field strengths must be expected. Efforts have therefore been made to include the spatial non-uniformity in calculations and to observe spatial structures, e.g. by more than one detector in the laser beam at different locations.

First we consider the case of a single transverse mode and the temporal stability of this mode (Section 10.1), then we consider the possibility of excitation of neighboring transverse modes (Section 10.2). In the latter case stationary or dynamical spatial patterns may arise as a consequence of the competition among transverse modes.

In Section 9.3.2 some comments on transverse multimode lasers were included.

10.1 Single Transverse Mode

The first example indicating that a transverse structure of the laser field might be important for the dynamics of lasers was given by Lugiato and Milani (1983 and 1985). It was shown that, if a fixed Gaussian laser mode transverse intensity distribution is assumed, in a homogeneously broadened laser there are no "bad-cavity" or "good-cavity" single-mode laser instabilities. In other words, neither single-mode instabilities (Chapter 5) nor instabilities involving multiple longitudinal modes (Chapter 9) occur.

It has been noted, of course, that the assumption of a fixed Gaussian beam profile is somewhat unrealistic. In a real laser one would more likely expect that the mode intensity profile or diameter would somehow intensity-dependently "breathe", that is, with higher laser power would increase and with lower laser power would decrease, or vice versa. This would mean that, e.g., the mode diameter would be another variable of the system.

In this respect, an experiment was performed (Lippi et al., 1987) in which spontaneous periodic pulsing of a CO_2 laser was observed. Figure 10.1.1 shows the experi-

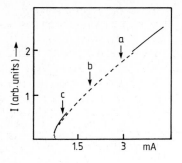

Fig. 10.1.1. Average output intensity of a CO_2 laser as a function of the discharge current. Solid lines, continuous output; broken line, pulsing. (From Lippi et al., 1987.)

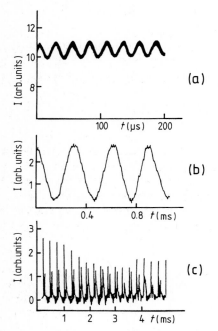

Fig. 10.1.2. Laser output intensity as a function of time at points (a), (b) and (c) marked in Fig. 10.1.1. (a) Oscillation amplitude = 10% of continuous laser output; (b) 60% of continuous output; (c) spiking with peak intensity ca. 30 times the cw laser output intensity. (From Lippi et al., 1987.)

mental findings. For low pumping (low laser discharge current) the laser operates continuously. The dashed line indicates periodic pulsing. At high pump rates the laser again emits continuously. The different types of pulses are marked in Fig. 10.1.1 as a, b and c and the corresponding pulse shapes are shown in Fig. 10.1.2; a is a small-amplitude sinusoidal modulation on a continuous background; b is a large amplitude similarly sinusoidal modulation; and c is a sharply spiked modulation. It is notable that there is hysteresis near the lower onset of pulsing between continuous and "spiky" emission. On going from a small to larger pump, the "spiky" pulsing thus appears abruptly with full amplitude.

A CO_2 laser is described by two equations only: the polarization relaxes so quickly that it follows the field adiabatically. These two equations allow no pulsing solutions, only damped relaxation oscillations. If one increases the phase space by adding the laser mode waist and the wavefront curvature as a dynamical variable, one finds a form of steady-state equations that suggest that this system may be described by the normal two laser equations with an intensity-dependent loss. Inserting the intensity dependence of the laser loss into the usual CO_2 laser equations, one finds the behavior pictured in Fig. 10.1.3, which closely resembles the experimental findings.

The conclusion to be drawn from this result is that lasers which in the plane-wave descriptions are predicted to be stable may actually be unstable, i.e. pulsing or chaotic, owing to the spatial structure of the pump and the generated field. The intensity dependence of the losses is easily understood in physical terms. If the mode "blows up" with increasing intensity, the losses due to diffraction, e.g. on the mirror edges, on diaphragms inside the resonator (such as the diameter of the discharge tube) increase.

10.2 Several Transverse Modes. Spatial Instabilities

So far, when speaking of instabilities we have thought of temporal behaviour, i.e. the onset of periodic or chaotic pulsing. On the other hand, one can of course also think of spatial instabilities, which means that the spatial structure of the laser mode changes at, e.g., a certain pump strength. One would, of course, then expect from a combination of spatial and temporal instabilities the phenomenon of "turbulence" as found in fluids or more generally in spatially extended systems with time delays in the coupling between spatially separated regions. We may imagine that the weather or socio-economic

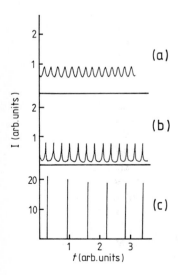

Fig. 10.1.3. Model calculation corresponding to Fig. 10.1.2. (From Lippi et al., 1987.)

processes show this kind of turbulent dynamics since the state of a volume element in these systems is determined not only by its own past but also by that of other volume elements in the past.

10.2.1 Theoretical Studies

Simple examples of spatial instabilities have been theoretically investigated by Lugiato and co-workers (Lugiato et al., 1987; Lugiato, Oldano et al., 1988; Lugiato, Prati et al., 1988; Lugiato, Oppo et al., 1988). A homogeneously broadened two-level laser was considered, with two types of resonators: a ring cavity with axial symmetry yielding uni-directional emission (Fig. 10.2.1) (Lugiato et al., 1987; Lugiato, Prati et al., 1988; Lugiato, Oppo et al., 1988), and a ring or Fabry-Pérot cavity which, in addition to the two end mirrors orthogonal to the longitudinal z-axis, has two lateral mirrors orthogonal to the x-axis (Lugiato, Oldano et al., 1988). Most of the results are qualitatively similar in both cases, so only the first is described here.

The distinctive feature of the studies mentioned is that several physical and geometrical features that are absent by definition in the plane-wave theory used in preceding chapters have been considered: diffraction effects caused by the finite transverse cross-section of the field and by its radial variations of amplitude and phase, wavefront curvature induced by the spherical mirrors and transverse and longitudinal gain variations caused by the pump mechanism. This leads to the following set of generalized Maxwell-Bloch coupled equations (Lugiato et al., 1987; Lugiato, Prati et al., 1988):

$$\frac{\partial F}{\partial \eta} + \frac{1}{v}\frac{\partial F}{\partial \tau} = \frac{i}{4}\left[\frac{\partial^2 F}{\partial \varrho^2} + \frac{1}{\varrho}\frac{\partial F}{\partial \varrho}\right] - \alpha L P$$

$$\frac{\partial P}{\partial \tau} = -[FD + (1 + i\delta_{AC})P] \qquad (10.2.1)$$

$$\frac{\partial D}{\partial \tau} = -\gamma\left[-\frac{1}{2}(F^*P + FP^*) + D - \chi(\varrho,\eta)\right]$$

where $F(r,z,t)$, $P(r,z,t)$ and $D(r,z,t)$ represent the normalized slowly varying envelopes of the electric field, atomic polarization and population difference, respectively. The radial and longitudinal coordinates r and z (Fig. 10.2.1) and the time t are related to ϱ, η and τ, respectively, by

$$\varrho = (\pi/L\lambda_0)^{1/2}r, \quad \eta = Z/L, \quad \tau = \gamma_\perp t \qquad (10.2.2)$$

where λ_0 denotes the wavelength of a cavity mode of frequency ω_0 taken as the reference frequency and $\gamma_\perp$ is the atomic linewidth. The parameter α is the gain coefficient per unit length; $v = c/L\gamma_\perp$, δ_{AC} is the difference between the atomic transition frequency and the cavity frequency ω_0, measured in units of $\gamma_\perp$; $\gamma = \gamma_\parallel/\gamma_\perp$ ($\gamma_\parallel$ is the damping rate of D) and $\chi(\varrho,\eta)$ is the unsaturated population inversion created by the pump mechanism.

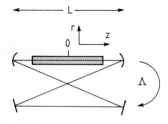

Fig. 10.2.1. Schematic representation of the ring resonator. The two spherical mirrors have identical radii of curvature R_0, the same power reflectivity R and are separated by a distance L. The full length of the resonator is Λ and the active medium length is L_A. (From Lugiato et al., 1987.)

Equations (10.2.1) were obtained in the paraxial approximation and under the simplifying assumption of cylindrical symmetry for the field and atomic variables. They are to be compared with the standard set (9.1.3) which corresponds to the plane-wave Maxwell-Bloch equations for the case of multiple longitudinal modes. The main difference is that an additional term proportional to the transverse Laplacian of F:

$$\nabla^2_\perp F = \left(\frac{\partial^2}{\partial r^2} + \frac{1}{r}\frac{\partial}{\partial r} \right) F \qquad (10.2.3)$$

appears in the first of eqns. (10.2.1). This term comes from the wave equation and accounts for diffraction effects in an axially symmetric geometry. Another difference between eqns. (10.2.1) and (9.1.3) consists in the absence of a term of the type $-kF$ describing cavity losses in the first of eqns. (10.2.1). This comes from the fact that Lugiato and co-workers (1987, 1988) found it more appropriate for the numerical solution of eqns. (10.2.1) to describe field losses by imposing boundary conditions requiring that the electric field at position $\eta = 0$ matches the field that has reached the position $\eta = \Lambda/L$ after propagating one full trip around the cavity.

For solving eqns. (10.2.1), the field may be expanded in a series of radial cavity modes:

$$F(\varrho,\eta,\tau) = e^{-i\delta\Omega\tau}\bar{F}(\varrho,\eta,\tau) = e^{-i\delta\Omega\tau}\sum_P A_p(\varrho,\eta)f_p(\eta,\tau) \qquad (10.2.4)$$

with the Gauss-Laguerre mode functions $A_p(\varrho,\eta)$ discussed by Ru et al. (1987), and $\delta\Omega$ is the offset between the unknown operating laser frequency and ω_0 in units of $\gamma_\perp$. By defining

$$P(\varrho,\eta,\tau) = e^{-i\delta\Omega\tau}\bar{P}(\varrho,\eta,\tau) \qquad (10.2.5)$$

substituting eqns. (10.2.4) and (10.2.5) into eqns. (10.2.1) and making use of the orthogonality and completeness relations of the functions $A_p(\varrho,\eta)$ with respect to the radial coordinate ϱ, one obtains the system

$$\frac{\partial f_p}{\partial \eta} + \frac{1}{v}\frac{\partial f_p}{\partial \tau} = \mathrm{i}\frac{\delta\Omega}{v}f_p - \alpha L \int_0^\infty \mathrm{d}\varrho\varrho\, A_p^* \bar{P}$$

$$\frac{\partial \bar{P}}{\partial \tau} = -[\bar{F}D + (1 + \mathrm{i}\varDelta)\bar{P}]$$

$$\frac{\partial D}{\partial \tau} = -\gamma\left[-\frac{1}{2}(\bar{F}^*\bar{P} + \bar{F}\bar{P}^*) + D - \chi\right]$$

(10.2.6)

where $\varDelta = \delta_{\mathrm{AC}} - \delta\Omega$. Equations (10.2.6) represent an infinite set of coupled equations. In practice, however, radial or "transverse" modes with p greater than 1 or 2 (let us assume that $p = 0$ corresponds to a TEM$_{00}$ mode and is excited) suffer significant diffraction losses, typically because of intracavity apertures, so that only a few transverse modes are expected to play a significant role in a real laser.

Until now eqns. (10.2.6) have only been solved under several simplifying conditions. One of them corresponds to the "extended uniform-field limit", which is defined by the constraints

$$\alpha L \to 0, \quad T \to 0 \quad \alpha L_{\mathrm{A}}/T = 2C = \text{arbitrary}$$

$$|\delta_p - 2\pi m| = O(1) \quad \text{for all } p \neq 0 \text{ and } m = 0 \pm 1 \pm 2 \ldots$$

(10.2.7)

where T is the transmittivity of the curved mirrors, $\delta_p = (\omega_p - \omega_0)\varLambda/c$ (ω_p is the frequency of the transverse mode p) and $2C$ is the gain parameter. The last constraint in eqns. (10.2.7) implies that the frequency separation between adjacent radial modes should be of the same order of magnitude as the frequency spacing between longitudinal resonances, and simultaneously excludes degeneracies between transverse and longitudinal modes. In the extended uniform-field limit, the mode amplitudes f_p are independent of η, and one of the steady-state solutions is dominated by the $p = 0$ mode ($f_0 = O(1)$ and $f_{p\neq0} = O(T)$), so that the cavity field configuration is of the TEM$_{00}$ type (Lugiato, Prati et al., 1988). The detuning parameter $\varDelta$ obeys the usual mode-pulling equation:

$$\varDelta = \frac{\delta_{\mathrm{AC}}}{1 + \kappa/\gamma_\perp}$$

(10.2.8)

where $\kappa = cT/\varLambda$ is the cavity linewidth.

Lugiato, Oppo et al. (1988) (and Lugiato, Oldano et al. (1988) with the four mirror-cavity) considered different conditions. The last constraint in eqns. (10.2.7) was replaced with the following:

$$\gamma_\perp \varLambda/c \ll 1 \quad (\kappa/\gamma_\perp \text{ arbitrary})$$

$$\delta_p \ll 1.$$

(10.2.9)

The first expression in eqns. (10.2.9), together with the first ones in eqns. (10.2.7), ensure that the dynamics of the system are governed by the single longitudinal mode that is nearest to the atomic line (resonant mode) as all other longitudinal modes lie outside the atomic bandwidth. The last condition in eqns. (10.2.9) means that the transverse modes fall in the vicinity of the resonant mode. This, together with the requirement $T \ll 1$, implies that the beam waist is nearly uniform along the atomic sample. Phenomenological damping coefficients κ_p were introduced (Lugiato, Oppo et al., 1988) to simulate the losses brought about by the finite diameter of the laser mirrors on each transverse mode amplitude.

We summarize below the main results obtained by Lugiato and co-workers with respect to the dynamic behavior. The four first points correspond to conditions where eqns. (10.2.7) and the "good-cavity" limit ($\kappa \ll \gamma_\perp$) apply and the longitudinal mode spacing is of the same order of magnitude as $\gamma_\perp$. The last point corresponds to conditions (10.2.9).

(i) On resonance ($\delta_{AC} = 0$), and in the presence of a longitudinal mode ($p = 0$), the adjacent radial modes $p \neq 0$ can become unstable. Figure 10.2.2 shows the ratio of the second laser threshold $2C^{(2)}$ to the first laser threshold $2C^{(1)}$ as a function of the transverse mode spacing $\delta\omega_1'$ (which is measured in units of $\gamma_\perp$). It is seen that for the first transverse mode $p = 1$ the instability threshold is only slightly higher than the first laser threshold. ($2C^{(2)}/2C^{(1)}$ is close to unity), mainly for small values of $\delta\omega_1'$ (which can be reached, for instance, by increasing the radius of curvature of the curved mirrors (Fig. 10.2.1)). The results are very sensitive to the ratio between the transverse size of the active volume (ϱ_0) and the beam cross-section: the instability is favored when the cavity is designed to operate with a large value of this ratio ($\gtrsim 3$).

(ii) Detuning of the laser influences the instability threshold, as is shown in the example in Fig. 10.2.3. In this case two indices (n and p) have been used to label each mode: n for the longitudinal character (this was omitted for simplicity in the previous figure and equations) and p for the transverse or radial character. One can observe that with increasing detuning the instability threshold increases. Figure 10.2.3 also shows that transverse modes belonging to the nearest longitudinal mode ($\bar{n} - 1$) can play an important role in the dyanmics of the system, primarily when their resonance

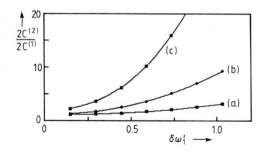

Fig. 10.2.2. Dependence of the second laser threshold $2C^{(2)}$ on the radial mode spacing $\delta\omega_1'$ for (a) $p = 1$, (b) $p = 2$ and (c) $p = 3$. As expected, the lowest radial modes are the most sensitive to the appearance of unstable behavior. The gain profile is given by $\chi(\varrho) = \exp(-\varrho^2/2\varrho_0^2)$. (From Lugiato, Prati et al., 1988.)

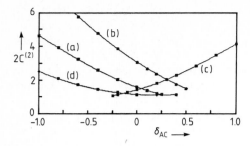

Fig. 10.2.3. Dependence of the second laser threshold $2C^{(2)}$ on the detuning parameter δ_{AC} for a number of radial modes, when the steady-state longitudinal mode $(\bar{n}, p = 0)$ is initially above threshold. (a) Corresponds to the radial mode $(\bar{n}, p = 1)$, (b) to $(\bar{n}, p = 2)$, (c) to $(\bar{n} - 1, p = 1)$ and (d) to $(\bar{n} - 1, p = 2)$. The pump profile is similar to that in Fig. 10.2.2. (From Lugiato, Prati et al., 1988.)

frequencies fall in the vicinity of the resonant mode (as occurs with the modes $(\bar{n} - 1, 1)$ and $(\bar{n} - 1, 2)$ in the example shown).

(iii) When a given pure longitudinal mode is excited, the neighboring pure longitudinal modes can be excited only under conditions of detuning, $\delta_{AC} \neq 0$ (Lugiato, Prati et al., 1988). As mentioned in Section 10.1, Lugiato and Milani (1983 and 1985) had already demonstrated earlier that the Risken-Nummedal instability (Chapter 9) disappears if every field mode has a TEM_{00} structure and the laser is operating on resonance, with a flat pump profile and a large Fresnel number $\pi W_0^2/\lambda_0 L_A \gg 1$ (W_0 is the beam waist) to ensure an uniform beam radius along the active region.

(iv) Unstable behaviors are found even in the rate-equation regime ($\partial P/\partial t = 0$) and in the limit when both atomic variables are eliminated adiabatically ($\partial P/\partial t = \partial D/\partial t = 0$). This result stands in striking contrast with the well known behavior of the plane-wave Maxwell-Bloch equations, which predict stable operation under these conditions.

(v) Under conditions (10.2.9), i.e. when the frequencies of the transverse modes are much closer to that of the resonant mode, the dynamical interaction of the transverse modes is enhanced. Instead of the usual competition (i.e. mode supression), in the "good-cavity" limit and operating on resonance one finds a range of values of the radial mode spacing for which strong temporal instabilities develop, leading to both periodic and chaotic spatio-temporal pulsations (Fig. 10.2.4.a), and an adjacent range where a new phenomenon called "cooperative frequency locking" (Lugiato, Oldano et al., 1988; Lugiato, Oppo et al., 1988) occurs. This gives rise to a steady-state regime where several radial modes show comparable amplitudes and their frequencies are pulled together, or frequency locked, to a common value (Fig. 10.2.4.b). Hysteresis effects between different steady-state configurations are also observed.

Recently, spontaneous breaking of the cylindrical symmetry in the transverse configuration has been predicted (Lugiato, Kaige et al., 1988). With the laser operating in a symmetrical single-mode configuration, an increase in the pump parameter is shown to produce a spatial instability, which causes the emergence of a radially asymmetrical steady-state pattern.

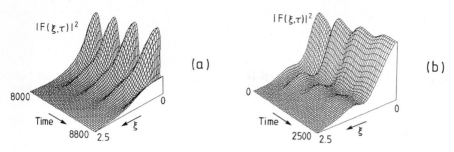

Fig. 10.2.4. Time-dependent profile of the radial intensity $|F(\xi, \tau)|^2$ as a function of the radial coordinate ξ (ξ is proportional to ϱ in our notation). (a) Corresponds to an unstable periodic configuration and (b) to a situation with a smaller radial mode spacing in which cooperative frequency locking is developing. (From Lugiato, Oppo et al., 1988.)

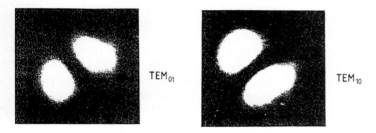

Fig. 10.2.5. Mode pattern of TEM_{01} and TEM_{10} modes.

As a final comment, we may point out that since the temporal dynamics depend on, e.g., the laser field strength, one may combine the spatial effects described here with the temporal chaos in lasers to give a very simple optical "turbulent" system which may be a model for other turbulent (space-time chaotic) systems that are too difficult to subject to experiment.

10.2.2 Experimental Results

The calculations first carried out for a waveguide resonator (Lugiato, Oldano et al., 1988) show a spatial instability manifesting itself very clearly in a breaking of the translational symmetry: at low excitation the laser field is homogeneous in the cross-section of the laser channel. After the fundamental and first transverse mode are locked, the laser output intensity has a periodic modulation across the laser channel cross-section. Such an experiment has not yet been conducted. For more common resonator structures, however, experimental verifications of this spatial laser instability exist.

The simplest transverse modes for laser resonators with spherical mirrors are the TEM_{01} and the TEM_{10} modes. In an empty resonator they have the same frequency, their intensity patterns (Fig. 10.2.5) being rotated by $90°$ with respect to each other. If the resonator contains some anisotropic element, e.g. Brewster angle windows, the frequency degeneracy of the two modes is lifted.

In the experiment described by Tamm (1988), a laser was forced to oscillate in modes with zero intensity at the center. This was achieved by a small absorbing dot placed at or near the center of the laser mode volume. The laser is then usually emitting in a TEM_{01} or TEM_{10} mode, or both. Compensating the anisotropy of the laser tube's Brewster angle windows is possible by displacing the absorber dot away from the center of the mode volume, thus effectively creating a second anisotropy. By moving the absorber dot, the frequency of the TEM_{01} and TEM_{10} modes can thus be made more or less equal. The frequency separation and the interaction of the two modes can be monitored by a photodiode recording the intensity at a particular location within the mode cross-section.

Figure 10.2.6 shows the spectra of the intensity monitored by the photodiode. At a large frequency separation, the two modes are essentially non-interacting and only a beat frequency appears (Fig. 10.2.6a). Decreasing the frequency difference increases the mode interaction, visible by the distortion of the beat — the harmonics of the beat frequency (Fig. 10.2.6b). For a small frequency spacing, the mode interaction becomes so strong that subharmonics of the beat frequency are generated (Fig. 10.2.6c) which eventually, as expected, terminate in chaotic emission (broad-band spectrum, Fig. 10.2.6d). A further reduction in the mode frequency separation then results in a sudden locking of the modes and the spectrum shows nothing except the zero-frequency marker

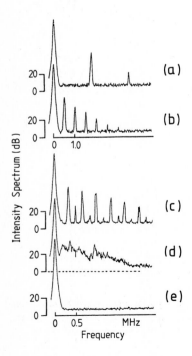

Fig. 10.2.6. Intensity spectra of laser oscillating in both TEM_{01} and TEM_{10} mode simultaneously. The modes have slightly different frequencies owing to anisotropy of the laser resonator. The mode frequency difference decreases from a to e. (From Tamm, 1988.)

10   Transverse Effects and Spatial Instabilities

(Fig. 10.2.6e). The mode pattern is then the well known "doughnut" (Fig. 10.2.7), also termed the TEM^*_{01} mode in the literature. The time picture of the intensities corresponding to the spectra in Fig. 10.2.6a−e are shown in Figs. 10.2.8a−d. Noteworthy is the behavior near the locking point. The pulses show relatively long "locked" conditions which at some point then abruptly "break". This behavior is reminiscent of the similar behavior of the laser with a field injected (Section 7.3).

The TEM^*_{01} mode has the interesting property of helical wavefronts: as is well known, the two bright spots of the TEM_{01} or TEM_{10} mode are $180°$ out of phase in the optical field. The formation of a bright ring without dark spokes is then possible only if the TEM_{01} and TEM_{10} modes are locked "in quadrature", that is, with a $90°$ phase difference between them.

Obviously there are two possibilities. The phase-locking angle can be $+90°$ or $-90°$, corresponding to a left- or a right-hand helix respectively. Consequently, the laser beyond the locking instability must be optically bistable. The optical bistability was tested experimentally (Tamm, 1989). Figure 10.2.9 shows a measured hysteresis curve.

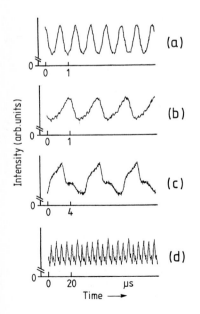

Fig. 10.2.7. TEM^*_{01} "Doughnut" mode arising from locking the TEM_{01} and TEM_{10} mode of Fig. 10.2.5 in phase quadrature.

(a)

(b)

(c)

(d)

Intensity (arb.units)

Time →

Fig. 10.2.8. Time pictures of the laser pulsing corresponding to the spectra in Fig. 10.2.6. (From Tamm, 1988.)

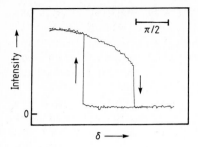

Fig. 10.2.9. Hysteresis curve obtained when switching between a left- and a right-hand helix field. The control parameter is the phase of the injected helix field. (From Tamm, 1989.)

The upper branch corresponds to emission on the left- and the lower branch to the right-hand helix mode. In the experiment it was verified that the switching between the two helices could be accomplished by injecting a suitable helix laser field into the laser.

This fundamental laser instability thus gives the laser a basic capability of "pattern recognition" and a "content-addressable memory". Further developments of this potentially very useful concept have to be awaited.

A second experiment conducted on a CO_2 laser used naturally non-degenerate modes: the fundamental TEM_{00} plus modes with higher indices. All modes in a spherical resonator become degenerate as the resonator approaches confocality. In the experiment of Tredicce et al. (1989), the effective curvature of the laser mirrors was changed by a telescope inside the resonator with variable distance between the two lenses. Note that a change in confocality does not change the laser frequency of the TEM_{00} mode in this arrangement. Figure 10.2.10 shows scans of the mode intensity profiles for telescope settings approaching confocality successively.

Figure 10.2.10a, which is far from confocality, shows only the TEM_{00} mode intensity profile. Changing the resonator towards confocality shows first a distortion of the mode profile (b). In (c) the beating with a new mode coming above threshold is seen, which locks with the first mode in (d), creating an even more complicated field structure. Moving further towards confocality shows the beating or interaction with a further mode coming above threshold (e), which again locks when confocality is further approached (f). The left-hand column shows the theoretical prediction (Lugiato, Oppo et al., 1988). The correspondence is satisfactory.

10.2.3   The Optical Vortex

A case where the realm of mode expansions, so familiar to lasers, has been abandoned to fully treat partial differential equations is the recently predicted occurrence of the analog of super-fluid vortices in optics (Coullet et al., 1989).

Equations like (10.2.1) were numerically integrated assuming as initial conditions, when the pump is zero, vacuum fluctuations everywhere in the laser cross section except at 4 points where the field was kept zero.

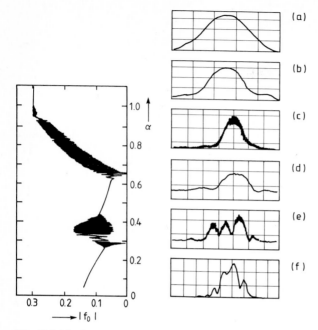

Fig. 10.2.10. Comparison of measured mode profiles of a CO_2 laser when the resonator is changed towards confocality (a)−(f). In locked cases the trace is sharp. In simultaneous oscillation cases of more than one frequency the beat can be seen on the mode profile. The left-hand side shows the theoretical prediction. f_0: laser field amplitude, α: parameter giving the distance from point of confocality of laser resonator. (From Tredicce et al., 1989.)

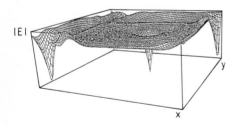

Fig. 10.2.11. Modulus of the field in the laser cross section. At the positions of the zeros of the initial conditions 4 vortices have developed after the laser pump is switched on. (From Coullet et al., 1989.)

In the calculations the pump was then switched on and the final solution iteratively obtained. Figure 10.2.11 shows the magnitude of the optical field in the laser cross section thus calculated.

Four points where the laser intensity is zero are apparent. A look at the polarization (or alternatively the field) Figure 10.2.12 shows the nature of the intensity zeros. From

Fig. 10.2.12. Equiphase lines of the solution whose field is shown in Fig. 10.2.11. 4 vortices are clearly recognizable. Phase difference between equiphase lines is $\pi/4$. (From Coullet et al., 1989.)

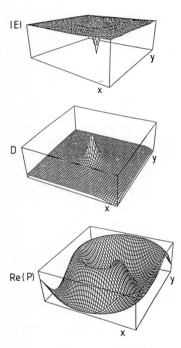

Fig. 10.2.13. Detailed view of the central vortex of Fig. 10.2.11 showing field E, inversion D, and the real part of the medium polarization. (From Coullet et al., 1989.)

the lines of constant phase it is seen that the zeros represent vortices. Figure 10.2.13 shows in more detail the properties of the central vortex of Fig. 10.2.11. Most surprising, at the vortex the population inversion is high above threshold. Even though this would appear to be an unstable situation, the solution is stable, which means, that the vortex itself prohibits a build up of emission at its location.

Although the phenomenon of vortices is predicted for a laser with a very large Fresnel number corresponding to a very large number of possible modes ($\sim 10^3$), it would appear that the basic elements of the vortex are already existing in systems with small numbers of modes. The helical wave emission laser has most of the properties of the vortex. In fact, it forms the central part of the vortex. However, in the helical wave laser, boundary conditions prevent laser emission in the centre, while there are no such boundary conditions for the vortex.

The vortex solution is coexisting with e.g. the constant intensity solution. While physically the existence of vortices links optics with fluids, solids and even cosmology (defects, magnetic monopoles, cosmic strings), the coexistence of the large variety of vortex solutions with e.g. the constant intensity solution indicates another time the capability of optical sysems to create spatially different coexisting states, a property of possible importance for future parallel data processing.

References

Chapter 1

Arecchi, F. T., Meucci, R., Puccioni, G.P., and Tredicce, J. R. (1982), *Phys. Rev. Lett.* **49**, 1217.

Brewer, R. G. (1977), *Physics Today* **30**, 50.

Brewer, R. G., and Shoemaker, R. L. (1972), *Phys. Rev.* **A6**, 2001.

Casperson, L. W. (1978), *IEEE Journ. Quant. El.* **QE−14**, 756.

Feigenbaum, M. J. (1978), *Journ. Stat. Phys.* **19**, 25.

Gibbs, H. M., Hopf, F. A., Kaplan, R. L., and Shoemaker, R. L. (1981), *Phys. Rev. Lett.* **46**, 474.

Grazyuk, A. Z., and Oraevskii, A. N. (1964) in Miles, P. A. (Ed.): *Quantum Electronics and Coherent Light.* Academic Press, New York.

Haken, H. (1975), *Phys. Lett.* **53A**, 77.

Libchaber, A., and Maurer, J. (1980), *J. Phys.* **41**, C3.

Lorenz, E. N. (1963), *J. Atm. Sci.* **20**, 130.

Maurer, J., and Libchaber, A. (1979), *J. Phys. Lett.* **40**, 419.

Nelson, D. F., and Boyle, W. S. (1962), *Appl. Opt.* **1**, 181.

Ruelle, D., and Takens, F. (1971), *Comm. Math. Phys.* **20**, 167.

Scholz, H., Yamada, T., Graham, R., and Brand, H. (1981), *Phys. Lett.* **82A**, 321.

Statz, H., and de Mars, G. (1960) in Townes, C. H. (Ed.): *Quantum Electronics.* Columbia University Press, New York, p. 530.

Treacy, E. B., and de Maria, A. J. (1969), *Phys. Lett.* **29A**, 369.

Weiss, C. O., Godone, A., and Olafsson, A. (1983), *Phys. Rev.* **A28**, 892.

Yamada, T., and Graham, R. (1980), *Phys. Rev. Lett.* **45**, 1322.

Zeghlade, H., and Mandel, P. (1985), *Journ. Opt. Soc. Amer.* **B2**, 18.

Chapter 2

Haken, H. (1975), *Phys. Lett.* **53 A**, 77.
Lorenz, E. N. (1963), *J. Atmos. Sci.* **20**, 130.
Sparrow, C. (1982): *The Lorenz Equations: Bifurcations, Chaos, and Strange Attractors.* Springer, Berlin-New York.

Chapter 3

Abraham, N. B., Lugiato, L. A., Mandel, P., Narducci, L. M., and Bandy, D. K. (1985), *J. Opt. Soc. Am.* **B2**, 35.
Arecchi, F. T., and Califano, A. (1986), *Europhysics Letters* **3**.
Bergé, P. Pomeau, Y. and Vidal, C. (1984): *L'ordre dans le chaos.* Hermann, Paris. Transl.: *Order within Chaos.* Wiley, New York 1986.
Eckmann, J.-P. (1981), "Roads to turbulence in dissipative dynamical systems", *Rev. Mod. Phys.* **53**, 643.
Eckmann, J.-P., and Ruelle, D. (1985), "Ergodic theory of chaos and strange attractors", *Rev. Mod. Phys.* **57**, 617.
Guckenheimer, J., and Holmes, P. (1983): *Nonlinear Oscillations, Dynamical Systems, and Bifurcations of Vector Fields.* Springer, New York.
Haken, H. (1983): *Advanced Synergetics.* Springer, Berlin.
Haken, H. (1985): *Synergetics. An Introduction.* Springer, Berlin, 3rd ed.
Hirsch, M. W., and Smale, S. (1974): *Differential Equations, Dynamical Systems and Linear Algebra.* Academic Press, New York.
Iooss, G., and Joseph, D. D. (1980): *Elementary Stability and Bifurcation Theory.* Springer, New York.
Ott, E. (1981), "Strange attractors and chaotic motions of dynamical systems", *Rev. Mod. Phys.* **53**, 655.
Schuster, H. G. (1988): *Deterministic Chaos. An Introduction.* VCH, Weinheim, 2nd ed.
Sparrow, C. (1982): *The Lorenz Equations: Bifurcations, Chaos, and Strange Attractors.* Springer, Berlin-New York.

Chapter 4

Benettin, G., Galgani, L., and Strelcyn, J. M. (1976), *Phys. Rev.* **A14**, 2338.
Bergé, P., Pomeau, Y., and Vidal, C. (1984): *L'ordre dans le chaos.* Hermann, Paris. Edition in English: *Order within Chaos.* Wiley, New York 1986.
Eckmann, J.-P., and Ruelle, D. (1985), "Ergodic theory of chaos and strange attractors", *Rev. Mod. Phys.* **57**, 617.
Grassberger, P., and Proccaccia, I. (1983a), *Physica* **D9**, 189.
Grassberger, P., and Proccaccia, I. (1983b), *Phys. Rev. Lett.* **50**, 346.
Grassberger, P., and Proccaccia, I. (1983c), *Phys. Rev.* **A28**, 2591.

Pesin, Y. B. (1977), *Usp. Mat. Nauk* **32**, 55; Engl. transl. (1977): *Math Surveys* **32**, 55.

Schuster, H. G. (1988): *Deterministic Chaos. An Introduction*. VCH, Weinheim, 2nd ed.

Shannon, C. E., and Weaver, W. (1949). *The Mathematical Theory of Information*. University of Illinois Press, Urbana.

Wolf, A., Swift, J. B., Swinney, H. L., and Vastano, J. A. (1985), *Physica* **D16**, 285.

Chapter 5

Abraham, N. B., Chyba, T., Coleman, M., Gioggia, R. S., Halas, N. J., Hoffer, L. M., Liu, S. N., Maeda, M., and Wesson, J. C. (1983) in Harvey, J. D., and Walls, D. F. (Eds.): *Laser Physics*. (Lecture Notes in Physics 182). Springer, Berlin.

Abraham, N. B., Lugiato, L. A., Mandel, P., Narducci, L. M., and Bandy, D. K. (1985), *J. Opt. Soc. Am.* **B2**, 35.

Ackerhalt, J. R., Milonni, P. W., and Shih, M.-L. (1985), *Phys. Rev.* **128**, 205.

Bandy, D. K., Narducci, L. M., Lugiato, L. A., and Abraham, N. B. (1985), *J. Opt. Soc. Am.* **B2**, 56.

Casperson, L. W. (1978), *IEEE J. Quant. El.* **QE14**, 756.

Casperson, L. W. (1980), *Phys. Rev.* **A21**, 911. Also in Harvey, J. D., and Walls, D. F. (Eds.): *Laser Physics*. Springer, Berlin 1983.

Casperson, L. W. (1981), *Phys. Rev.* **A23**, 248. Also in Harvey, J. D., and Walls, D. F. (Eds.): *Laser Physics*. Springer, Berlin 1983.

Casperson, L. W. (1983) in Harvey, J. D., and Walls, D. F. (Eds.): *Laser Physics*. Springer, Berlin, p. 88.

Casperson, L. W. (1985a), *J. Opt. Soc. Am.* **B**, 62 and 73.

Casperson, L. W. (1985b), *J. Opt. Soc. Am.* **B2**, 993.

Casperson, L. W. (1988), *J. Opt. Soc. Am.* **B5**, 958 and 970.

Corbalan, R., Laguarta, F., Pujol, J., and Vilaseca, R. (1989), *Opt. Comm.* **71**, 290.

Dupertuis, J. A., Salomaa, R. E., and Siegrist, M. R. (1987), *IEEE Journ. Quant. El.* **QE23**, 1217.

Englund, J. C. (1986), *Phys. Rev.* **A33**, 3606.

Gioggia, R. S., and Abraham, N. B. (1983a), *Phys. Rev. Lett.* **51**, 650.

Gioggia, R. S., and Abraham, N. B. (1983b), *Opt. Comm.* **47**, 278.

Gioggia, R. S., Abraham, N. B., Lange, W., Tarroja, M. F. H., and Wesson, J. C. (1988), *J. Opt. Soc. Am.* **B5**, 992.

Graham, R., and Cho, Y. (1983), *Opt. Commun.* **47**, 52.

Haken, H. (1966), *Z. Physik* **190**, 327.

Haken, H. (1985a): *Light* (Vol. 2). North-Holland, Amsterdam.

Haken, H. (1985b): *Synergetics: An Introduction*. Springer, Berlin, 3rd ed.

Hendow, S. T., and Sargent III, M. (1982), *Opt. Commun.* **40**, 385, and **43**, 59.

Hendow, S. T., and Sargent III, M. (1985), *J. Opt. Soc. Am.* **B2**, 84.

Hillman, L. W., Boyd, R. W., and Stroud Jr., C. R. (1982), *Opt. Lett.* **7**, 426.

Hoffer, J. M., Chyba, T. H., and Abraham, N. B. (1985), *Journ. Opt. Soc. Am.* **B2**, 102.

Khandokin, P. A., Khanin, Ya. I., and Koryukin, I. V. (1988), *Opt. Comm.* **65**, 367.

Klische, W., Weiss, C. O., and Wellegehausen, B. (1989), *Phys. Rev.* **A39**, 919.

Laguarta, F., Pujol, J., Vilaseca, R., and Corbalán, R. (1988), *J. Physique* **49**, C2, 409. Also in Firth, W., Peyghambarian, N., and Tallet, A. (Eds.): *Proceedings of Optical Bistability IV*. Les Editions de Physique, Les Ulis 1988.

Lawandy, N. M., and Plant, D. V. (1986), *Opt. Comm.* **59**, 55.

Lawandy, N. M., and Rabinovich, W. S. (1986) in Boyd, R. W., Raymer, M. G., and Narducci, L. M. (Eds.): *Optical Instabilities*. Cambridge University Press, Cambridge, p. 240.

Lawandy, N. M., Plant, D. V., and Lee, K. (1987), *Phys. Rev.* **A36**, 3253.

Lawandy, N. M., and Lee, K. (1987), *Opt. Comm.* **61**, 137.

Lugiato, L. A., Narducci, L. M., Bandy, D. K., and Abraham, N. B. (1985), *Opt. Commun.* **46**, 115.

Lugiato, L. A., Narducci, L. M., and Squicciarini, M. F. (1986), *Phys. Rev.* **A34**, 3101.

Maeda, M., and Abraham, N. B. (1982), *Phys. Rev.* **A26**, 3395.

Mandel, P. (1983), *Opt. Commun.* **44**, 269 and 400.

Mandel, P. (1985), *J. Opt. Soc. Am.* **B2**, 112.

Mandel, P., and Zeghlache, H. (1983), *Opt. Commun.* **47**, 146−150.

Mandel, P., and Erneux, T. (1984), *Phys. Rev. Lett.* **53**, 1818.

Moloney, J. V., Uppal, J. S., and Harrison, R. G. (1986), *Phys. Rev. Lett.* **59**, 2868.

Narducci, L. M., Sadiky, H., Lugiato, L. A., and Abraham, N. B. (1985), *Opt. Comm.* **55**, 370.

Puccioni, G. P., Tratnik, M. V., Sipe, J. E., and Oppo, G. L. (1987), *Opt. Lett.* **12**, 242.

Risken, H., Schmid, C., and Weidlich, W. (1966), *Z. Physik* **194**, 337.

Ryan, J. C., and Lawandy, N. M. (1987), *Opt. Comm.* **64**, 59.

Sargent III, M., Scully, M. O., and Lamb, W. E. (1974): *Laser Physics*. Addison-Wesley, Reading.

Sargent III, M., Zubairy, M. S., and de Martini, F. (1983), *Opt. Lett.* **8**, 76.

Shih, M.-L., Milonni, P. W., and Ackerhalt, J. R. (1985), *J. Opt. Soc. Am.* **B2**, 130.

Siemsen, K., Reid, J., and Weiss, C. O. (1987), *Opt. Comm.* **64**, 54.

Sparrow, C. (1982): *The Lorenz Equations: Bifurcations, Chaos and Strange Attractors*. Springer, Berlin.

Tarroja, M. F. H., Abraham, N. B., Bandy, D. K., and Narducci, L. M. (1986), *Phys. Rev.* **A34**, 3148.

Uppal, J. S., Harrison, R. G., and Moloney, J. V. (1987), *Phys. Rev.* **A36**, 4823.

Weiss, C. O., Klische, W., Ering, P. S., and Cooper, M. (1985), *Opt. Comm.* **52**, 405.

Weiss, C. O., Abraham, N. B., and Hübner, U. (1988), *Phys. Rev. Lett.* **61**, 1587.

Weiss, C. O., Abraham, N. B., and Hübner, U. (1989), *Phys. Rev.* **A40**, 6354.

Wu, T. Q., and Weiss, C. O. (1987), *Opt. Commun.* **61**, 337.

Zeghlache, H., and Mandel, P. (1985), *J. Opt. Soc. Am.* **B2**, 18.

Chapter 6

Antoranz, J. C., Bonilla, L. L., Gea, J., and Velarde, M. G. (1982), *Phys. Rev. Lett.* **49**, 35.

Antoranz, J. C., and Rubio, M. A. (1988), *J. Opt. Soc. Am.* **B5**, 1070−1073.

Arimondo, E., Casagrande, F., Lugiato, L. A., and Glorieux, P. (1983), *Appl. Phys.* **B 30**, 57 – 77.

Arimondo, E., Bootz, P., Glorieux, P., and Menchi, E. (1985), *J. Opt. Soc. Am.* **B 2**, 193 – 201.

Arimondo, E., de Tomasi, F., Zambon, B., Papoff, F., and Hennequin, D. (1988), *J. Physique* **49**, C 2, 123 – 126 (Proceedings of *Optical Bistability* IV).

Bekkali, A., Papoff, F., Dangoisse, D., and Glorieux, P. (1988), *J. Physique* **49**, C 2, 349 – 354 (Proceedings of *Optical Bistability* IV).

Burak, I., Houston, P. L., Sutton, D. G., and Steinfeld, J. I. (1971), *IEEE Journ. Quant. El.* **QE 7**, 73 – 82.

Chyba, D. E. (1988), *J. Physique* **49**, C 2, 367 – 370 (Proceedings of *Optical Bistability* IV).

Chyba, D. E., Abraham, N. B., and Albano, A. M. (1987), *Phys. Rev.* **A 35**, 2936.

Dangoisse, D., Bekkali, A., Papoff, F., and Glorieux, P. (1988), *Europhysics Letters* **6**, 335.

Dangoisse, D., and Glorieux, P. (to be published), *J. Opt. Soc. Am.* **B**.

Dembinski, S. T., Kossakowski, A., Peplowski, P., Lugiato, L. A., and Mandel, P. (1978), *Phys. Lett.* **68 A**, 20 – 22.

Dupré, J., Meyer, F., and Meyer, C. (1975), *Rev. Phys. Appl.* (Paris) **10**, 285 – 293.

Englund, J. C. (1988), *J. Opt. Soc. Am.* **B 5**, 1033.

Erneux, T. (1988), *J. Opt. Soc. Am.* **B 5**, 1063 – 1069.

Erneux, T., and Mandel, P. (1981), *Z. Phys.* **B 44**, 353 – 374.

Garcia – Fernández, P., and Velarde, M. G. (1988), *J. Opt. Soc. Am.* **B 5**, 1074 – 1076.

Haken, H. (1985): *Synergetics: An Introduction.* Springer, Berlin, 3rd ed.

Hennequin, D., de Tomasi, F., Zambon, B., and Arimondo, E. (1988), *Phys. Rev.* **A 37**, 2243 – 2246.

Hennequin, D., Bekkali, A., Dangoisse, D., and Glorieux, P., *personal communication*.

Heppner, J., Solajić, Z. and Merkle, G. (1984), *Appl. Phys.* **B 35**, 77 – 82.

Jaques, A., and Glorieux, P. (1982), *Opt. Commun.* **40**, 455 – 460.

Kawaguchi, H., and Iwane, G. (1981), *Electron. Lett.* **17**, 167.

Kawaguchi, H. (1982), *IEE Proc.* **129**, Pt. I, 141.

Kawaguchi, H. (1984), *Appl. Phys. Lett.* **45**, 1264.

Kusnetzow, M., Tsang, D. Z., Walpole, J. N., Lian, Z. L., and Ippen, E. P. (1986) in Boyd, R. W., Raymer, M. G., and Narducci, L. M. (Eds.): *Optical Instabilities.* Cambridge University Press, Cambridge, p. 281.

Lugiato, L. A., Mandel, P., Dembinski, S. T., and Kossakowski, A. (1978), *Phys. Rev.* **A 18**, 238.

Mandel, P., and Erneux, T. (1984), *Phys. Rev.* **A 30**, 1893.

Mrugala, F., and Peplowski, P. (1980), *Z. Phys.* **38 B**, 359.

Salomaa, R., and Stenholm. S. (1973), *Phys. Rev.* **A 8**, 2695 – 2726.

Shilnikow, L. P. (1965), *Sov. Math. Dokl.* **6**, 163.

Tachikawa, M., Tanii, K., Kajita, M., and Shimizu, T. (1986), *Appl. Phys.* **B 39**, 83 – 90.

Tachikawa, M., Tanii, K., and Shimizu, T. (1987), *J. Opt. Soc. Am.* **B 4**, 387 – 395.

Tachikawa, M., Tanii, K., and Shimizu, T. (1988), *J. Opt. Soc. Am.* **B 5**, 1077 – 1082.

Tachikawa, N., Hong, F.-L., Tanii, K., and Shimizu, T. (1988a), *Phys. Rev. Lett.* **60**, 2266.

Tachikawa, M., Hong, F.-L., Tanii, K., and Shimizu, T. (1988b), *Phys. Rev. Lett.* **61**, 1042.

Tanii, K., Tachikawa, M., Kajita, M., and Shimizu, T. (1988), *J. Opt. Soc. Am.* **B5**, 24 – 28.

Tomasi, F., Hennequin, D., Zambon, B., and Arimondo, E. (1989), *J. Opt. Soc. Am.* **B6**, 45.

Ueda, K., and Shimizu, F. O. (1984), *Jpn. J. Appl. Phys.* **23**, 1038 – 1044.

Velarde, M. G., and Antoranz, J. C. (1986) in Boyd, R. W., Raymer, M. G., and Narducci, L. M.: *Optical Instabilities*. Cambridge University Press, Cambridge.

Winful, H. G., Shen, Y. C., and Liu, J. M. (1986), *Appl. Phys. Lett.* **48**, 616.

Chapter 7, Section 1

Arecchi, F. T., Meucci, R., Puccioni, G., and Tredicce, J. (1982), *Phys. Rev. Lett.* **49**, 1217.

Arecchi, F. T., Lippi, G. L., and Puccioni, G. P. (1984), *Opt. Comm.* **51**, 308.

Arecchi, Gadomski, W., Meucci, R., and Roversi, J. A. (1988), *J. Physique* **49** (6), C2, 363. Also in Firth, W., Peyghambarian, N., and Tallet, A. (Eds.) (1988): Proceedings *Optical Bistability* IV. Les Editions de Physique, Les Ulis.

Biswas, D. J., Dev, V., Chatterjee, U. K. (1987), *Phys. Rev.* **A35**, 450.

Brun, E., Derighetti, B., Meier, D., Holzner, R., and Ravani, M. (1985), *Journ. Opt. Soc. Am.* **B2**, 156.

Chen, Y. C., Winful, H. G., and Liu, J. M. (1985), *Appl. Phys. Lett.* **47**, 208.

Dangoisse, D., Glorieux, P., and Hennequin, D. (1986), *Phys. Rev. Lett.* **57**, 2657.

Dangoisse, D., Glorieux, P., and Hennequin, D. (1987), *Phys. Rev.* **A36**, 4775.

Erneux, T., Baer, S. M., and Mandel, P. (1987), *Phys. Rev.* **A35**, 1165.

Glorieux, P. (1987), *J. Physique* **48**, C7, 433.

Grothe, H., Harth, W., and Russer, P. (1976), *Electon. Lett.* 12, 522.

Haken, H., and Sauermann, H. (1963), *Z. Phys.* **176**, 47.

Hennequin, D., Bekkali, A., Dangoisse, D., and Glorieux, P., *personal communication.*

Hori, Y., Serizawa, H., and Sato, H. (1988), *J. Opt. Soc. Am.* **B5**, 1128.

Ivanov, D. V., Khanin, Ya. I., Matorin, I. I., and Pikovsky, A. S. (1982), *Phys. Lett.* **89A**, 229.

Kawaguchi, H. (1984), *Appl. Phys. Lett.* **45**, 1264.

Khandokhin, P. A., and Khanin, Y. I. (1984), *Sov. J. Quant. El.* **14**, 1004.

Klische, W., Telle, H. R., and Weiss, C. O. (1984), *Opt. Lett.* **9**, 561.

Kubodera, K., and Otsuka, K. (1981), *IEEE J. Quant. El.* **QE17**, 1139.

Lauterborn, W., and Steinhoff, R. (1988), *J. Opt. Soc. Am.* **B5**, 1097.

Lee, C. H., Yoon, T. H., and Shin, S.-Y. (1985), *Appl. Phys. Lett.* **46**, 95.

Mandel, P., and Erneux, T. (1984), *Phys. Rev. Lett.* **53**, 1818.

Mandel, P., Nardone, P., and Erneux, T. (1988), *J. Opt. Soc. Am.* **5**, 1113.

Midavaine, T., Dangoisse, D., and Glorieux, P. (1985), *Phys. Rev. Lett.* **55**, 1989.

Otsuka, K. (1978), *IEEE J. Quant. El.* **QE14**, 49.

Ohtsu, M., Teramachi, Y., Miyazake, M., and Weiss, C. O. (to be published), *Jap. Journ. Appl. Phys.*

Phillips, M. W., Gong, H., Ferguson, A.I., and Hanna, C. D. (1987), *Opt. Comm.* **61**, 215.

Polushkin, N. I., Khandokhin, P. A., and Khanin, Y. I. (1983), *Sov. J. Quant. El.* **13**, 956.

Scholz, H., Yamada, T., Brand, H., and Graham, R. (1981), *Phys. Lett.* **82 A**, 321.

Schuster, H. G. (1988): *Deterministic chaos.* VCH, Weinheim, 2nd ed.

Shore, K. A. (1988), *J. Opt. Soc. Am.* **B5**, 1211.

Statz, H., Luck, C., Shater, C., and Ciftan, M. (1961) in Singer, J. R. (Ed.): *Advances in Quantum Electronics.* Columbia University Press, New York.

Tang, C. L. (1963), *J. Appl. Phys.* **34**, 2935.

Tredicce, J. R., Abraham, N. B., Puccioni, G. P., and Arecchi, F. T. (1985), *Opt. Commun.* **55**, 131.

Tredicce, J. R., Arecchi, F. T., Lippi, G. L., and Puccioni, G. P. (1985), *J. Opt. Soc. Am.* **B2**, 173.

Tredicce, J. R., Arecchi, F. T., Puccioni, G. P., Poggi, A., and Gadomski, W. (1986), *Phys. Rev.* **A34** 2073.

Weiss, C. O., and Cho, Y. (unpublished work).

Wiesenfeld, K., and McNamara, B. (1986), *Phys. Rev.* **A33**, 629.

Yamada, T., and Graham, R. (1980), *Phys. Rev. Lett.* **45**, 1322.

Chapter 7, Section 2

Arecchi, F. T., Degiorgio, V., and Querzola, B. (1967), *Phys. Rev. Lett.* **19**, 168.

Arecchi, F. T., and Politi, A. (1980), *Phys. Rev. Lett.* **45**, 1215.

Arecchi, F. T., Gadomski, W., and Meucci, R. (1986), *Phys. Rev.* **A34**, 1617.

Arecchi, F. T., Meucci, R., and Gadomski, W. (1987), *Phys. Rev. Lett.* **58**, 2205.

Arecchi, F. T., Gadomski, W., Lapucci, A., Mancini, H., Meucci, R., and Roversi, J. A. (1988), *J. Opt. Soc. Am.* **B5**, 1153.

Arecchi, F. T., Lapucci, A., Meucci, R., Roversi, J. A., and Coullet, P. H. (1988), *Europhys. Lett.* **6**, 677.

Agrawal, G. P., and Henry, C. H. (1988), *IEEE J. Quant. El.* **QE24**, 134.

Chen, L.-X., Li, Ch.-F., Hu, Q.-S., Li, J.-F., and Abraham, N. B. (1988), *J. Opt. Soc. Am.* **B5**, 1160.

Haake, F. (1978), *Phys. Rev. Lett.* **41**, 1685.

Henry, C. H. (1982), *IEEE J. Quant. El.* **QE18**, 259.

Henry, C. H. (1983), *IEEE J. Quant. El.* **QE19**, 1391.

Henry, C. H., and Kazarinov, R. F. (1986), *IEEE J. Quant. El.* **QE22**, 294.

Lang, R., and Kobayashi, K. (1980), *IEEE J. Quant. El.* **QE16**, 347.

Li, H., and Abraham, N. B. (1988), *Appl. Phys. Lett.* **53**, 2257.

Li, H., and Abraham, N. B. (1989), IEEE *J. Quant. El.* **QE 25**, 1782.

Spano, P., Piazzolla, S., and Tamburrini, M. (1984), *IEEE J. Quant. El.* **QE 20**, 350.
Swetits, J. J., and Buoncristiani, A. M. (1988), *Phys. Rev.* **A 38**, 5430.
Telle, H. R., and Li, H. (1989), *IEEE J. Quant. El.* **QE 25**, 257.

Chapter 7, Section 3

Arecchi, F. T., Lippi, G. L., Puccioni, G. P., and Tredicce, J. R. (1984a), *Opt. Commun.* **51**, 308.
Arecchi, F. T., Lippi, G. L., Puccioni, G. P., and Tredicce, J. R. (1984b) in Mandel, L., and Wolf, E. (Eds.): *Coherence and Quantum Optics V*. Plenum, New York.
Bandy, D. K., Narducci, L. M., and Lugiato, L. A. (1985), *J. Opt. Soc. Am.* **B 2**, 148.
Boulnois, J. L., van Lerberghe, A., Cottin, P., Arecchi, F. T., and Puccioni, G. P. (1986), *Opt. Comm.* **58**, 124.
Brun, E., Derighetti, B., Meier, D., Holzner, R., and Ravani, M. (1985), *J. Opt. Soc. Am.* **B 2**, 156.
Brun, E., Derighetti, B, Ravani, D., Broggi, G., Meier, P., and Stopp, R. (1986), *Physica Scripta* **T 13**, 119.
Holzner, R., Derighetti, B., Ravani, M., and Brun, E. (1987), *Phys. Rev.* **A 36**, 1280.
Lugiato, L. A., Narducci, L. M., Bandy, D. K., and Pennise, C. A. (1983), *Opt. Commun.* **46**, 64.
Morozov, V. N. (1988), *J. Opt. Soc. Am.* **B 5**, 909.
Otsuka, K. (1985), *J. Opt. Soc. Am.* **B 2**, 168.
Otsuka, K., and Iwamura, H. (1983), *Phys. Rev.* **A 28**, 3153.
Otsuka, K., and Kawaguchi, H. (1984a), *Phys. Rev.* **A 29**, 2953.
Otsuka, K., and Kawaguchi, H. (1984b), *Phys. Rev.* **A 30**, 1575.
Scholz, H., Yamada, T., Brand, H., and Graham, R. (1981), *Phys. Lett.* **A 82**, 321.
Spencer, M. B., and Lamb Jr., W. E. (1972), *Phys. Rev.* **A 5**, 884.
Tredicce, J. R., Arecchi, F. T., Lippi, G. L., and Puccioni, G. P. (1985), *J. Opt. Soc. Am.* **B 2**, 173.
Weiss, C. O., Bava, E., de Marchi, A., and Godone, A. (1980), *IEEE J. Quant. El.* **QE 16**, 498.
Yamada, T., and Graham, R. (1980), *Phys. Rev. Lett.* **45**, 1322.

Chapter 8

Abraham, N. B., Mandel, P., and Casperson, L. W. (1983), *IEEE J. Quant. El.* **QE 19**, 1724.
Abraham, N. B., Dangoisse, D., Glorieux, P., and Mandel, P. (1985), *J. Opt. Soc. Am.* **B 2**, 23.
Corbalán, R., Laguarta, F., Pujol, J., and Vilaseca, R. (1989), *Opt. Commun.* **71**, 290.
Davis, B. W., Vass, A., Pidgeon, C. R., Allan, G. R. (1981), *Opt. Comm.* **37**, 303.
Dupertuis, M. A., Salomaa, R. R. E., and Siegrist, M. R. (1986), *Opt. Commun.* **57**, 410.

Forysiak, W., Harrison, R. G., and Moloney, J. V. (1989), *Phys. Rev.* **A 39**, 421.

Harrison, R. G., Biswas, A. J. (1985), *Phys. Rev. Lett.* **55**, 63.

Heppner, J., and Hübner, U. (1980), *IEEE J. Quant. El.* **QE 16**, 1093.

Hogenboom, E., Klische, W., Weiss, C. O., and Godone, A. (1985), *Phys. Rev. Lett.* **55**, 2571.

Khandokhin, P. A., Khanin, Ya. I., and Koryukin, I. V. (1988), *Opt. Comm.* **65**, 367.

Klische, W., and Weiss, C. O. (1985), *Phys. Rev.* **A 31**, 4049.

Klische, W., Weiss, C. O., and Wellegehausen, B. (1989), *Phys. Rev.* **A 39**, 919.

Laguarta, F., Pujol, J., Vilaseca, R., and Corbalán, R. (1988), *J. Physique* **49**, C2, 409. Also in Firth, W., Peyghambarian, N., and Tallet, A. (Eds.): *Proceedings of Optical Bistability IV*. Les Editions de Physique, Les Ulis 1988.

Laguarta, F., Pujol, J., Vilaseca, R., and Corbalán, R. (1989), in Pesquera, L., and Bermejo, F. J. (Eds.): *Dynamics of Non-Linear Optical Systems*. World Scientific, Singapore, p. 187.

Lawandy, N. M., and Koepf, G. (1980), *IEEE J. Quant. El.* **QE 16**, 701.

Lawandy, N. M., and Plant, D. V. (1986), *Opt. Commun.* **59**, 55.

Lawandy, N. M., and Ryan, J. C. (1987), *Opt. Commun.* **63**, 53.

Lefebvre, M., Dangoisse, D., and Glorieux, P. (1984), *Phys. Rev.* **A 29**, 758.

Marx, R., Abdul-Halim, I., Weiss, C. O., Hübner, U., Heppner, J., Ni, Y. C., and Willenberg, G. D. (1981), *IEEE J. Quant. El.* **QE 17**, 1123.

Mehendale, J. C., and Harrison, R. G. (1986a), *Phys. Rev.* **A 34**, 1613.

Mehendale, J. C., and Harrison, R. G. (1986b), *Opt. Comm.* **60**, 257.

Moloney, J. V., Uppal, J. S., and Harrison, R. G. (1987), *Phys. Rev. Lett.* **59**, 2868.

Narducci, L. M., Sakiky, H., Lugiato, L. A., and Abraham, N. B. (1985), *Opt. Comm.* **55**, 370.

Pujol, J., Laguarta, F., Vilaseca, R., and Corbalán, R. (1988a), *J. Opt. Soc. Am.* **B 5**, 1004 – 1010.

Pujol, J., Laguarta, F., Corbalán, R., and Vilaseca, R. (1988b), *J. Phys. D: Appl. Phys.* **21**, 180.

Roldán, E., de Valcárcel, G. J., Vilaseca, R., Silva, F., Pujol, J., Corbalán, R., and Laguarta F., (1989), *Opt. Comm.* **73**, 506 .

Ryan, J. C., and Lawandy, N. M. (1987), *Opt. Commun.* **64**, 54.

Ryan, J. C., and Lawandy, N. M. (1987), *Phys. Rev.* **A 30**, 4823.

Siemsen, K., Reid, J., and Weiss, C. O. (1987), *Opt. Comm.* **64**, 54.

Sparrow, C. (1982): *The Lorenz Equations: Bifurcations, Chaos, and Strange Attractors*. Springer, Berlin.

Uppal, J. S., Harrison, R. G., and Moloney, J. V. (1987), *Phys. Rev.* **A 36**, 4823.

Weiss, C. O., unpublished work.

Weiss, C. O., Abraham, N. B., and Hübner, U. (1988), *Phys. Rev. Lett.* **61**, 1587.

Weiss, C. O., and Brock, J. (1986), *Phys. Rev. Lett.* **57**, 2804.

Weiss, C. O., and Klische, W. (1984), *Opt. Comm.* **51**, 47.

Weiss, C. O., and Klische, W., unpublished work.

Wellegehausen, B. (1979), *IEEE J. Quant. El.* **QE 15**, 1108. Also in Button, K. J., Inguscio, M., and Strumia, F. (Eds.) (1984): *Reviews of Infrared and Millimeter Waves*, Vol. 2. *Optically Pumped Far Infrared Lasers*.

Wellegehausen, B., Shadin, S., Friede, D., and Welling, H. (1977), *Appl. Phys.* **13**, 97.
Wu, T. Q., and Weiss, C. O. (1987), *Opt. Commun.* **61**, 337.
Wu, T. Q., Schrödev, V., Saad, F. I., and Klische, W., to be published (Infrared Physics, 1989).
Wu, X.-G., and Mandel, P. (1985), *J. Opt. Soc. Am.* **B3**, 724.

Chapter 9

Abraham, N. B., and Weiss, C. O. (1988), *Opt. Comm.* **68**, 437.
Ackerhalt, J. R., Milonni, P. W., and Shih, M.-L. (1985), *Phys. Rep.* **128**, 205.
Atmanspacher, H., and Scheingraber, H. (1986), *Phys. Rev.* **A34**, 253.
Baer, T. (1986), *J. Opt. Soc. Am.* **B3**, 1175.
Bezaeva, L. G., Kaptzov, L. N., and Sharipov, I. Z. (1985), *Kvantovaya Elektron.* **12**, 1743.
Biswas, D. J., and Harrison, R. G. (1985), *Phys. Rev.* **A32**, 3835.
Brunner, W., Fischer, R., and Paul, H. (1988), *J. Opt. Soc. Am.* **B5**, 1139.
Cho, Y., and Umeda, T. (1986), *Opt. Comm.* **59**, 131.
Danileiko, M. V., Kravchik, A. L., Nechiporenko, V. N., Tselinko, A. M., and Yatsenko, L. P. (1986), *Sov. J. Quant. El.* **16**, 1420.
French, P. M. W., Gomes, A. S. L., Gouveia-Neto, A. S., and Taylor, J. R. (1986), *Opt. Comm.* **60**, 389.
Fu, H., and Haken, H. (1987), *Phys. Rev.* **A36**, 4802.
Fu, H., and Haken, H. (1988), *J. Opt. Soc. Am.* **B5**, 899.
Glas, D., Müller, R., and Klehr, A. (1983), *Opt. Comm.* **47**, 297.
Golyaev, U. D., and Lantratov, S. V. (1983), *Kvantovaya Elektron.* **10**, 925.
Graham, R., and Haken, H. (1968), *Z. Phys.* **213**, 420.
Hauck, R., Hollinger, F., and Weber, M. (1983), *Opt. Comm.* **47**, 141.
Haken, H. (1985): *Light.* North-Holland, Amsterdam, Vol. 2.
Hendow, S. T., and Sargent III, M. (1982), *Opt. Comm.* **40**, 385 and **43**, 59.
Hillman, L. W., Boyd, R. W., and Stroud Jr., C. R. (1982), *Opt. Lett.* **7**, 426.
Hillman, L. W., Krasinski, J., Boyd, R. W., and Stroud Jr., C. R. (1984), *Phys. Rev. Lett.* **52**, 1605.
Hillman, L. W., and Koch, K. (1986) in Boyd, R. W., Raymer, M. G., and Narducci, L. M.: *Optical Instabilities.* Cambridge University Press, Cambridge.
Hoffer, L. M., Lippi, G. L., Abraham, N. B., and Mandel, P. (1988), *Opt. Comm.* **66**, 219.
Jacobs, S. F., Sargent III, M., Scully, M. O., Simpson, J., Sanders, V., and Killpatrick, J. E. (Eds.) (1984): *Physics of Optical Ring Gyros.* The International Society for Optical Engineering (SPIE), Bellingham.
Khandokhin, P. A., and Khanin, Ya. I. (1985), *J. Opt. Soc. Am.* **B2**, 226.
Khanin, Ya. I. (1988), *J. Opt. Soc. Am.* **B5**, 889.
Klische, W., and Weiss, C. O. (1985), *Phys. Rev.* **A31**, 4049.
Kuznetsov, M., Tsang, D. S., Walpole, J. N., Liau, Z., Ippen, E. P. (1987), *Appl. Phys. Lett.* **51**, 895.

Lett, P., and Mandel, L. (1985), *J. Opt. Soc. Am.* **B2**, 1615.

Lippi, G. L., Tredice, J. R., Arecchi, F. T., and Abraham, N. B. (1985), *Opt. Comm.* **53**, 129.

Lugiato, L. A., Narducci, L. M., Eschenazi, E. V., Bandy, D. K., and Abraham, N. B. (1985), *Phys. Rev.* **A32**, 1563.

Lugiato, L. A., and Narducci, L. M. (1985), Phys. Rev. **A32**, 1576.

MacFarlane, D. L., Casperson, L. W., and Tovar, A. A. (1988), *J. Opt. Soc. Am.* **B5**, 1144.

Mandel, P., and Agrawal, G. P. (1982), *Opt. Comm.* **42**, 269.

Mandel, P., and Abraham, N. B. (1984), *Opt. Comm.* **51**, 87.

Mandel, L., Roy, R., and Singh, S. (1984) in Bowden, C. W., Ciftan, M., and Robl, H. R. (Eds.) (1984): *Optical Bistability*. Plenum Press, New York.

Melnikov, L. A., Rabinovich, E. M., and Tuchin, V. V. (1988), *J. Opt. Soc. Am.* **B5**, 1134.

Mukai, T., and Otsuka, K. (1985), *Phys. Rev. Lett.* **55**, 1711.

Narducci, L. M., Tredicce, J. R., Lugiato, L. A., Abraham, N. B., and Bandy, D. K. (1985), *Phys. Rev.* **A32**, 1588.

Narducci, L. M., Tredicce, J. R., Lugiato, L. A., Abraham, N. B., and Bandy, D. K. (1986), *Phys. Rev.* **A33**, 1842.

New, G. H. C. (1983), *Rev. Prog. Phys.* **46**, 877.

O'Gorman, J., Phelau, P., McInerney, J., and Heffernan, D. (1988), *J. Opt. Soc. Am.* **B5**, 1105.

Otsuka, K., and Kawaguchi, H. (1984), *Phys. Rev.* **A30**, 1575.

Risken, H., and Nummedal, K. (1968a), *J. Appl. Phys.* **39**, 4662.

Risken, H., and Nummedal, K. (1968b), *Phys. Lett.* **A26**, 275.

Sargent III, M., Zubairy, M. S., and de Martini, F. (1983), *Opt. Lett.* **8**, 76.

Shih, M.-L., and Milonni, P. W. (1984), *Opt. Comm.* **49**, 155.

Weiss, C. O., Godone, A., and Olafsson, A. (1983), *Phys. Rev.* **A28**, 892.

Zeghlache, H., Mandel, P., Abraham, N. B., Hoffer, L. M., Lippi, G. L., and Mellow, T. (1988), *Phys. Rev.* **A37**, 470.

Chapter 10

Coullet, P., Gil, L., and Rocca, F. (1989), *Opt. Comm.* (in print).

Lippi, G. L., Abraham, N. B., Puccioni, G. D., Arecchi, F. T., and Tredicce, J. R. (1987), *Phys. Rev.* **A35**, 3978.

Lugiato, L. A., and Milani, H. (1983), *Opt. Comm.* **46**, 57.

Lugiato, L. A., and Milani, H. (1985), *J. Opt. Soc. Am.* **B2**, 15.

Lugiato, L. A., Kaige, W., Narducci, L. M., Oppo, G. L., Pernigo, M. A., Tredicce, J. R., Prati, F., and Bandy, D. K. (1988), European Quantum Electronics Conference, Hannover, Sept. 1988.

Lugiato, L. A., Oldano, C., and Narducci, L. M. (1988), *J. Opt. Soc. Am.* **B5**, 879.

Lugiato, L. A., Oppo, G.-L., Pernigo, M. A., Tredicce, J. R., Narducci, L. M., and Bandy, D. K. (1988), *Opt. Comm.* **68**, 63.

Lugiato, L. A. Prati, F., Bandy, D. K., Narducci, L. M., Ru, P., and Tredicce, J. R. (1987), *Opt. Comm.* **64**, 167.

Lugiato, L. A., Prati, F., Narducci, L. M., Ru, P., Tredicce, J. R., and Bandy, D. K. (1988), *Phys. Rev.* **A 37**, 3847.

Ru, P., Narducci, L. M., Tredicce, J. R., Bandy, D. K., and Lugiato, L. A. (1987), *Opt. Comm.* **63**, 310.

Tamm, Chr. (1988), *Phys. Rev.* **A 38**, 5960.

Tamm, Chr. (1989), *Phys. Rev. Lett.*, submitted March 1989.

Tredicce, J. R., Quel, E. J., Ghazzawi, A. M., Green, C., Pernigo, M. A., Narducci, L. M., and Lugiato, L. A. (1989), *Phys. Rev. Lett.* **62**, 1274.

Subject Index

Chaotic attractor 50f., 59f., 70, 89, 147f., 156, 192ff.
—, Lorenz 24, 73, 89, 94, 101f., 179, 194
Chaotic behavior, characterization 67ff., 109
Chaotic emission 8
—, *see also* Pulsing, chaotic
Chaotic laser dynamics 12
Characteristic exponents 43, 46
Class A lasers 14, 87, 172
Class B lasers 14, 87, 137, 146, 168, 234
—, single-mode 159
Class C lasers 14, 87
Closed orbits 42, 52, 58
—, unstable 42, 52
Codimension-two bifurcation point 127, 198
Coexistence of multiple states *see* Multistability
Coherence of light for pumping 181
Coherent effects 14, 110, 187f., 191ff., 201
Coherent pump effects 191ff.
Complexity, degrees 67, 139
Computer experiment 67f., 70f.
Confocality 252f.
Conservative system 10, 142
Continous wave (cw) emission 3
Convection flows in fluids 17
Cooperative frequency locking 248
Co-propagating longitudinal modes 209ff.
Cosmic strings 255
CO_2 (gas) lasers 9, 92, 198f., 207
—, bidirectional 218ff.
—, modulated 137ff.
—, multimode 230f.
—, with electrical feedback 158ff.
—, with injected optical fields 168ff.
—, with saturable absorbers 119ff., 129ff.
Counter-propagating modes 218ff.
Coverage 72f.
Crises 59, 141, 147f., 156f.
Current modulation 153f.

Degree of chaoticity 76
Degrees of complexity 67, 139
Degrees of freedom 9, 12, 15, 139, 159
—, reducing the number of 36ff.
Delay, the choice of 77
Density matrix 106
Deterministic chaos 51, 59, 67, 214
—, vs. noise 75

Detuning 91, 99, 107f., 113f., 117, 131, 142, 149, 162, 168f., 170, 172ff., 186, 195, 200, 213
Devil's staircase 134
Differential equations non-linear 35f.
Diffraction effects 244
Dimensions
—, calculation 78f., 238f.
—, correlation dimension 69, 73f.
—, embedding 68, 77f.
—, fractal 13, 22, 73, 105, 238
—, generalized 69, 73f., 79f.
—, Hausdorff 22, 50, 69, 72f., 80
—, information dimension 69, 74
—, Kolmogorov 73
Diode lasers 154
—, difference to other lasers 234
—, self-pulsing 133, 239f.
—, with external reflectors or resonators 234ff.
—, *see also* Semiconductor lasers
Dispersion effect 110
—, anomalous 174
Dissipative pertubation 142
Dissipative systems 10, 36, 47
Distributed feedback mode 92
Doppler broadening 192, 194
Doppler effects 13, 85, 93, 111, 176, 193, 204
Doppler shift 187, 201
Doughnut mode 251
Dye lasers 13, 215ff., 240
Dynamical stability 39, 54
Dynamic behavior
—, non-linear systems 12f., 35f.
—, physical quantities for the characterization 69ff.
—, practical examples 77ff.

Electric conductivity 82
Electron-phonon interactions 127
Electro-optic modulator 146, 159, 167
Entropy 69, 74, 149
Ergodic theory 69
Evolution of physical systems 36ff.
—, *see also* Dynamic behavior
Evolution equation 36, 86f.

Fabry-Pérot cavity 111
Fabry-Pérot interferometer, non-linear 174

Fabry-Pérot resonator 91, 181, 198, 204
Far-infrared gas lasers 19, 181, 183f., 188ff.
−, see also Ammonia laser
Feedback lasers 158ff.
−, electrical feedback 165ff.
−, optical feedback 163ff.
Feedback loop 159, 162f.
Feedback phase 165
Feigenbaum scenario 60ff., 140, 193
−, see also Period doubling
Feigenbaum type of onset of chaos 9
Fiber laser 152
Field phase 195
First laser threshold 8, 11, 26, 92, 96f., 247
Fixed-point attractor 48, 89
Fixed points 24, 39, 44, 54, 131
−, global stable manifold 41
−, local stable manifold 40, 42
−, marginally stable 40
−, stable, focus 39f.
−, −, node 39f., 56
−, unstable 40, 52, 161
Floquet matrix 44, 46
Floquet multiplier 44, 56f., 62
Floquet theory 43, 46, 56, 140
Flow patterns 11
Fluctuation enhancement 161
Fluid flows 10
Fluids 15, 255
Four-level model 121f.
Fractal attractor 50f., 60
Fractal boundaries 48
Fractal dimension 13, 22, 73, 105, 238
−, see also Hausdorff dimension
Fractal objects 22
−, examples 22f., 25
Free induction decay 14
Frequency-locking 169
Frequency noise 165
Fresnel number 230, 154

Gas lasers 13, 81ff., 105, 168ff., 181, 202f.
−, see also CO_2 (gas) lasers
Gain, unsaturated 201
Gain-linked dispersive optical system 174
Gain modulation 149ff.
Gain-splitting 183, 187, 197
Gain variations 244
Gaussian beam profile 241

Generalized dimensions 69, 73f., 79f.
Giant pulses 170ff., 177
Global bifurcation 52, 57f.
Global stable manifold 41
Golden (and silver) mean 134f.
Good-cavity limit 210, 214, 247f.
Growth law 78

Hard-mode excitations 53, 89, 171
Hausdorff dimension 22, 50, 69, 72f., 80
−, see also Fractal dimension
Helical wavefronts 251, 255
He-Ne laser 9, 111, 118
−, bidirectional 226f.
−, multimode 231ff.
Heteroclinic orbit 41, 52, 57
Heterodyne measurements 98, 115ff., 195, 198, 205f., 218
−, experimental set-up 116
Heterodyne power spectra 109f.
He-Xe laser 111ff., 117
Highly confined system 10
Hole burning 110f.
−, spatial 7, 13, 174, 176, 209, 217f.
Homoclinic cycle 52, 126, 130f.
Homoclinic explosion 58
Homoclinic orbit 19, 21, 41, 52f., 57f., 126, 131, 161
Homodyne and heterodyne spectra 117
Homogenous broadening factors 85
Homogeneously broadened lasers 7, 13, 185, 191, 197
−, experimental results 92ff.
−, general equations 87f.
−, non-resonant case 90f.
−, other related cases 91f.
−, resonant case 88ff.
−, with ring resonator 209, 218
Hopf bifurcation 55ff., 59f., 145, 161, 166, 198, 200
−, subcritical 55f., 89f., 179
−, −, first observation 178
−, supercritical 55, 90, 108, 125f., 165, 178
Hyperchaos 53, 127
Hyper-torus 50, 57, 59
Hysteresis 89, 146f., 158, 238, 242, 248, 251f.

Lorenz model 17, 32, 36f., 41, 64, 77ff., 95ff., 103ff., 165, 179, 181, 188, 191, 193, 196, 205
—, determination of dimension 78f.
Lorenz equations 17ff., 87, 89, 95, 99ff., 186, 194
—, fixed points 24ff.
—, normal (stable) solutions 18f.
—, periodic windows 26
—, strange solutions 20ff.
Lyapunov exponents 46f., 53, 72, 131, 149f.
—, definition 46
—, distinguishing between periodic and chaotic attractors 70
—, Kaplan-Yorke conjecture 75
—, K-entropy 76
Lyapunov time 103

Magnetic monopoles 255
Magnetic sublevel degeneracy 187
Manifolds 40ff.
Map, definition 30, 37
—, describing a dynamical system 51
Master lasers 168
Maxwell-Bloch (-type) equations, generalized 244ff.
—, plane wave 86, 91, 248
Maxwellian distribution 105
Metastable chaos 6, 19, 22
Mid-infrared lasers, continuous wave (cw) 202f., 206f.
—, pulsed 206f.
Mode diameter 241
—, number 15
Mode-lock destruction 240
Mode-locked pulsing 111
Mode-locking, active 209, 240
—, longitudinal 6f.
—, passive 209, 213, 215, 238f.
Mode pattern 251
Mode splitting 90, 110
—, spontaneous or pump-induced 198, 200f.
Modes, co-propagating longitudinal 209ff.
—, counter-propagating 218ff.
—, single transverse 241ff.
—, transverse 229ff., 243ff.
Modulated lasers
—, chaotic dynamics 132ff.
—, CO_2 gas lasers 137ff., 146ff.
—, experiments 146ff.

—, NMR laser 155ff., 177ff.
—, semiconductor diode lasers 144ff., 153f.
—, solid state lasers 144ff.
—, theoretical description 137ff.
Modulation of laser parameters 137
—, gain 149ff.
—, pump rate 140, 142f.
—, relaxation oscillation frequency 140
—, resonator frequency 143f.
—, resonator loss 139ff., 146ff.
Modulator, acousto-optic 202, 208
Multimode instabilities 210ff., 214
—, dye lasers 13, 215, 217
—, link to single-mode instabilities 213
Multimode lasers 6f., 209ff.
—, mode-mode interaction leading to chaos 217
—, routes to chaos 231, 233
—, transverse 230ff.
—, with intracavity frequency doubling 227ff.
Multistability 52, 147, 153, 156, 239f.
—, generalized 53, 66
Multi-rhythmicity 53

Nd: YAG laser 152f.
—, bidirectional, fluctuation spectra 225
—, multimode 230
—, with an internal frequency doubler 227f.
Near-infrared lasers, continous wave (cw) 203ff.
NMR lasers 137, 146, 155ff., 177ff.
Noise 75, 78f.
—, of frequency fluctuations 165
—, influence on scenarios 26, 65f.
—, reduction 158
Noise sources 164
Noisy periods 61
Noisy period doublings 29, 153ff.
Non-linear maps 30, 37f.
Non-linear dynamical systems 12f., 34, 67
—, time evolution 35f.
Non-linear dynamic behavior 12f., 15
—, asymptotic behavior 47ff.
—, examples 17ff.
—, mathematical concepts 9, 35ff.
Non-linear oscillators, damped 146
Non-wandering set 52
Non-Lorenzian behavior 191
N_2O laser 132